Drilling and Producing Offshore

Drilling and Producing Offshore

R. Stewart Hall

PennWell Books
PennWell Publishing Company
Tulsa, Oklahoma

Library of Congress Cataloging in Publication Data
Main entry under title:

Drilling and producing offshore.

1. Oil well drilling, Submarine. 2. Oil
wells. I. Hall, R. Stewart.
TN871.3.L37 1983 622'.338 82-18956
ISBN 0-87814-213-4

PennWell Publishing Company
1421 South Sheridan Road/P. O. Box 1260
Tulsa, Oklahoma 74101

Library of Congress cataloging in publication data

Printed in the United States of America

2 3 4 5 87 86 85

Contents

Foreword

Offshore drilling and production have always represented a highly complex and technically challenging mosaic of operations. With the phenomenal surge in offshore activities witnessed in the late 1970s and early 1980s and the resultant introduction of innovative equipment and techniques, the demand has grown for more and more qualified, highly specialized individuals to design and execute the myriad phases of drilling and production.

However, with this specialization it has become increasingly difficult for any one individual to gain a general overview of the full scope of activities involved and how the new equipment and techniques fit into the overall picture.

The purpose of this volume is to attempt to present the reader with technically detailed examinations of specialized, state-of-the-art designs, equipment, and techniques and to present a comprehensive overview of the great spectrum of offshore drilling and production operations.

I would like to express my sincere gratitude to the individual authors who contributed their hard work and expertise. This book is a tribute to their dedication to meeting the continuing challenges of the offshore industry.

I would also like to acknowledge the generous cooperation of the many companies and individuals who provided their technical assistance and counsel. These farsighted, industry-minded experts are always willing to work together in a joint effort to present examples of the advancing technology found in the offshore energy industry.

R. Stewart Hall

Introduction

With world demand for oil constantly increasing, drilling and production activity in recent years has been concentrated more and more in once-inaccessible offshore areas. Offshore wells are no longer novelties: the Gulf of Mexico has long been liberally sprinkled with them. Gulf offshore drilling and production began in the marshes, estuaries, and bays and later spread to deeper waters as new discoveries were made and as equipment and techniques were improved. Today more than half of the world's offshore wells are in the Gulf of Mexico. The skills and technology developed there have provided the foundation for development of offshore oil fields around the world.

The basic requirement for offshore operations is a structure from which to work and on which to place necessary equipment. Early estuary operations utilized the techniques and equipment of land operations and crude wooden platforms. As drilling moved farther offshore, economic, environmental, and technical requirements caused the operating techniques and equipment—as well as the supporting structures—to become more sophisticated and the industry responded with innovative answers.

In this reference work, we have assumed that the first three steps of the normal sequence of events have occurred (Fig. I–1). The exploratory stage is performed to define the shape, size, and estimated content of recoverable hydrocarbons in the reservoir. Exploratory drilling utilizes barges, jackup rigs, submersibles, semisubmersibles, drillships, and other types of units, depending on water depth, weather conditions, distance offshore, and the availability of the type of rig required. These rigs will drill from four to ten wells in the reservoir for definition and will then move off to another location or block—as close as 20 miles or as far away as 5,000 miles.

The rigs are designed for efficiency and mobility. Most of the fleet is made up of jackup and semisubmersible rigs because of their adaptability to the changing conditions worldwide. They are not designed to stay on location and produce the reservoir; rather, they perform the important stages of res-

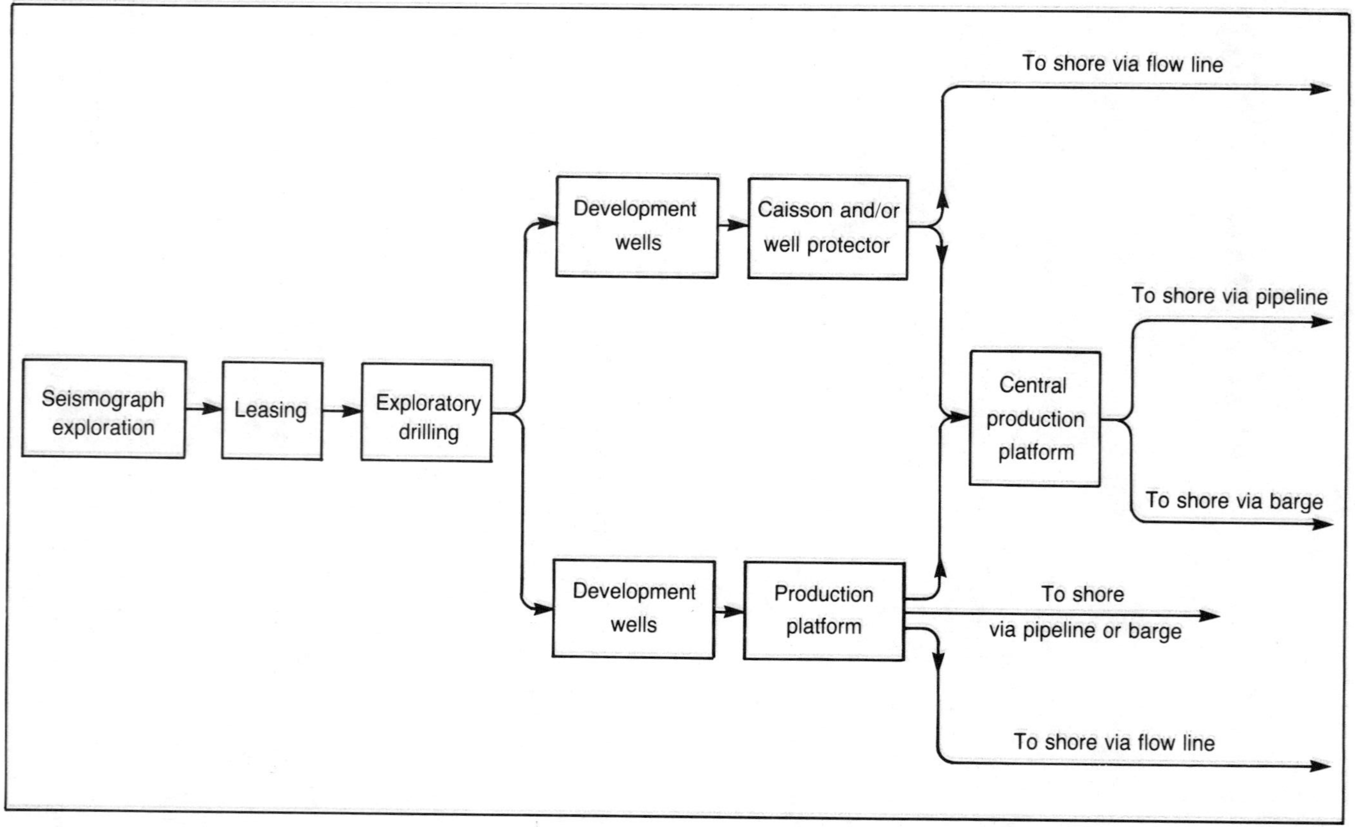

FIG. i–1 Scheme of methods for exploring, developing, and producing hydrocarbon deposits.

ervoir definition upon which the design criteria are based in order to build a drilling production structure. The types of platform structures used in the exploratory drilling stage are chronicled and discussed in a companion volume, *The Technology of Offshore Drilling, Completion, and Production* (PennWell Books, 1976).

Tomorrow's developments in recovering hydrocarbons offshore will no doubt bring many changes in the production sciences, and designers must plan for new dramatic systems to place upon these structures in order to produce a far greater percentage of the estimated reservoir. By today's calculations, we are recovering an average of 40% of known reserves. With new technology and with world demand growing, this percentage will improve dramatically within the next decade.

I

Development Drilling Structures

Michael W. Ridge
R. Stewart Hall and Associates

1

Template, Concrete Gravity, and Other Platforms

The term *platform* encompasses many kinds of structures. The structure may be very small over a single well in shallow water, or it may be very large over dozens of wells in deep water. Such a large structure is probably remotely located, necessitating living quarters, communication facilities, and transportation installations such as a helicopter pad. It is a virtual village with space at a premium and safety a major concern. The platform is a permanent fixture that will remain in place for the producing life of the wells, and it must be able to withstand any environmental or operational condition that may arise.

The design of every offshore platform—regardless of type and location throughout the world's waters—must meet four basic requirements: (1) the size must be adequate for the intended operations, (2) the structure must be capable of supporting the operational equipment loads and all other ancillary support facilities, (3) construction methods, both of fabrication and installation, must be practical, and (4) the cost must be reasonable.

The 1970s witnessed a remarkable surge in the overall numbers of offshore platforms throughout the world and saw the evolution of several new platform designs. Before 1970, virtually all platforms in use in the world's waters were of the piled, steel-jacket, template design (Fig. 1–1). Although there had been some research and development devoted to more innovative and exotic platform design, it was not until the 1970s that the actual testing, prototyping, and finally installation of several new types of platforms took place. So while the template platform remains the dominant structure in terms of overall numbers, record water depths, and production capabilities, the newer platform designs represent a growing and well-accepted segment of the offshore platform population.

Platform data summary	
Conductors	48
Environmental data	
Water depth	400 feet
Wave height	58 feet
Wind (one minute mean)	166 mph
Environmental loads	
Wave	6,100k - $6,500^k$
Wind	$1,200^k$ - $1,500^k$
Drilling loads	$12,000^k$ - $15,000^k$
Schedule	
Engineering	5 months
Fabrication	12 months
Installation	½ month
Tonnages	
Deck	$1,250^T$ - $1,600^T$
Jacket	$4,800^T$ - $5,200^T$
Piling	$3,200^T$ - $3,500^T$
Cost (millions of dollars)	
Engineering	0.15 - 0.2
Fabrication & materials	12.2 - 13.6
Installation	2.5 - 3.0
Total	14.85 - 16.8

FIG. 1–1 Drilling/production platform, double rig operation.

In addition to template platforms, some of the newer platform designs that will be discussed in this chapter include concrete gravity platforms, tension-leg platforms, drill-through-the-leg platforms, and the guyed-tower platform.

While each design type may contain unique and innovative features, every type still incorporates—in its own fashion—all of the basic requirements of an offshore platform: each provides a safe, economical, efficient structure for drilling and producing in the world's bountiful offshore oil and gas reserves.

It is impractical to design and construct the ultimate platform built for all contingencies and conditions. What must be designed is a structure that will reasonably withstand all known hazards in a specific locality and that will provide support for all operations for the life of the reservoir.

In most areas of the world, once the reservoir has been completed, the platform must be removed.

No two platforms are exactly alike, just as no two oil or gas wells are identical (Fig. 1–2). A few feet of horizontal difference or vertical height will dictate changes in the basic design. Considerable planning must be given to ensure the structure will provide sufficient area for both initial and future operational needs. During the production stages, the pay zone may vary a great deal during the life of the reservoir, necessitating changes in deck equipment and overall weight. Any modifications that may be anticipated and planned for will provide substantial savings in fabrication costs.

Platform designers must consider several criteria before selecting an appropriate structure. Some of these considerations include water depth, weather conditions, bottom conditions, size of the reservoir and anticipated production levels, the method of fabrication and installation, and overall cost factors.

Water depth has become an increasingly significant criterion of design as production has moved into deeper waters. While the definition of what is deep water has changed during the progress of the offshore industry, it traditionally has been considered as the depth just beyond the deepest existing platform. The designs of platforms for deep waters, while not totally unrelated to those for shallow waters, are dominated by factors that are less important in shallow-water platforms. Stresses due to wave dynamics and installation difficulties increase dramatically as deeper waters are encountered. Fatigue is a greater problem. The vastly increased weight in a deep-water platform

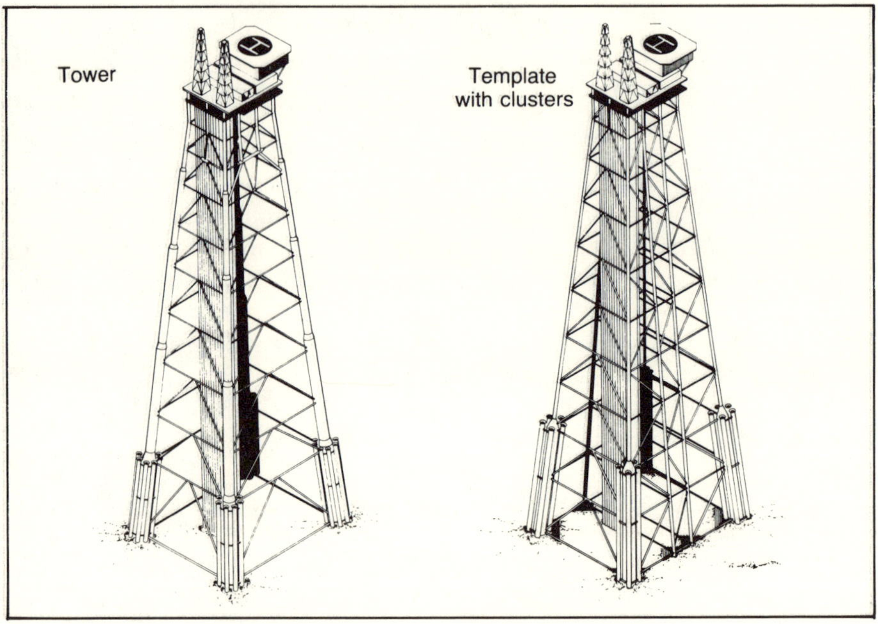

FIG. 1–2 Types of offshore platforms.

makes conventional installation techniques inadequate, and innovative concepts must be utilized. The number of wells per platform is likely to be increased in order to justify the large investment in a deep-water structure.

All structures are planned with a worst-case weather factor in mind. Normally this is a 100-year storm, meaning a storm with a probability of 1/100 of occurring in any year. Unlike ships, which can seek a safe harbor to wait out extreme weather, a drilling and production platform must bear the brunt of any violent forces nature may hurl against it. Hurricanes in the Gulf of Mexico, typhoons in Southeast Asia, and violent storms in the North Sea are severe weather problems that must be translated into major design criteria for a platform. Structural strength, platform foundation support, and height of working deck above mean water level are directly related to anticipated worst-case weather conditions.

Sea-bottom conditions must also be considered for design of the foundation system that anchors the platform to the sea bed. A hard, level, stable sea floor may be appropriate for gravity platforms, but soft, unstable sea bottoms may require extensive utilization of anchor piles driven as deep as 500 ft below the mud line. Subsea mudslides are encountered in delta areas near the mouth of some rivers such as the Mississippi. Here, enormous shifts of sea bottom may put intolerable loading stresses upon a platform base. Earthquakes in a particular locality or ice formations found in Arctic waters are additional environmental problems that must be acknowledged; appropriate design features must be incorporated to insure the safety of the platform workers.

Once a reservoir has been defined, its size and the anticipated production levels of hydrocarbons will influence the platform design. Is the reservoir large enough to justify a large, multiwell structure or will a smaller, single-well platform suffice? Will the production be offloaded onto tankers, pipelined to land facilties, or stored within the platform?

Because fabrication of a platform is performed onshore with installation taking place at sea, the construction concept must be considered in its design. Launching and installation load factors may become critical elements in platform design.

Finally, what are the economics of the entire structure? Construction costs vary tremendously, based on location and conditions. Since labor and material costs differ between countries and existing authorities regulate the designs, costs for similar structures can double or triple. A platform for the North Sea and another for the Gulf of Mexico—although nearly identical in appearance—can vary in costs by as much as a factor of three. The economics of leasing rights and the actual drilling and production expenses must also be considered with the platform costs. Especially for huge, deep-water platforms, final payout (the point at which the income from production equals the total platform investment costs) may take several years. In extreme cases it may never occur during the life of the reservoir. With the overall costs of developing some reservoirs that require the superplatforms already in the $2+ billion range, the question is whether the world demand for oil and gas will make it sufficiently profitable to justify further development of these reservoirs.

Considering all design criteria, a choice of platform type must be made. Throughout the history of the offshore industry for most applications, a pile-supported template platform design has been selected. These structures have survived many engineering evolutions and have withstood every environmental problem in the world's waters. With the state of technology today, they will no doubt continue to be the prevailing structure of choice for the foreseeable future.

Today's pile-supported template platforms are universally of the steel-jacket design. The base of the jacket is anchored to the sea floor with piles.

Depending upon bottom conditions and anticipated environmental loads, the piles are driven several hundred feet into the earth below the mud line. These pilings are driven either through the jacket legs themselves and grouted into place or they are driven through skirts or sleeves attached to the base of the platform (Fig. 1–3). The most common method of sinking the piles is by steam hammers operated from a derrick barge. Underwater pile-driving equipment has been successfully employed but only in unique deep-water situations.

From the sea floor, steel legs rise to above the water surface to support the deck assembly. The number and size of the legs and the structural reinforcement by crossmembers or braces are determined by previously discussed design criteria. The deck assembly consists of deck plates supported on a system of beams, trusses, or girders that distribute the operational loadings to the substructure legs through a tubular space frame.

Drilling is carried out through individual well conductors that provide protection against environmental forces to the surface casing.

Fabrication of template platforms takes place on land. Most large fabrication facilities specializing in offshore structures have the hoisting capabilities to lift virtually all sizes of jacket components presently involved in template platform construction. With two remarkable exceptions—the *Hondo* and the *Cognac* platforms, which will be discussed later—the entire jacket is assembled in one piece.

The assembled jacket is usually carried on a barge to the installation site and is lifted or launched into the water. In those cases where a deep-water jacket is too heavy to load onto a barge or when launching equipment is not available, a self-floating jacket may be utilized. Since flotation must be built into the jacket, this concept involves more steel and thus more complex and costly fabrication.

One of the most critical design criteria for the platform is this launching or towing operation. With the launch barges presently available to the industry, the water depth limit for one-piece launched jackets is about 950 ft.

The jacket, after launching or being towed to the site, must be rotated from horizontal to vertical and positioned on the sea floor. In shallow waters this traditionally is accomplished by a combination of lifting and flooding. But in deeper waters, because of the difference in response to hydrodynamic forces between the derrick barge and the jacket, controlled flooding alone must be used, requiring a great deal of sophistication in design and fabrication.

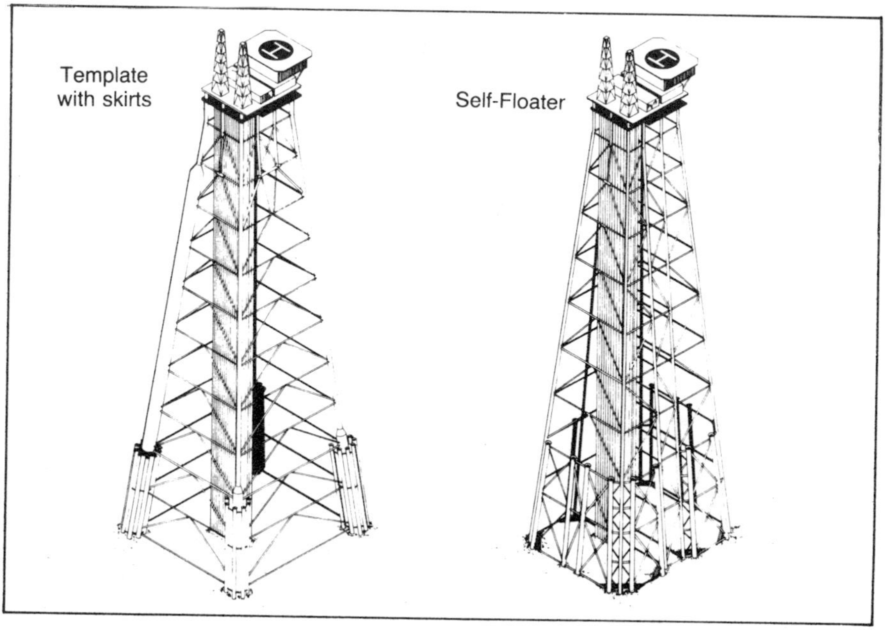

FIG. 1–3 Template with skirts (a) and self-floater (b).

Installation is completed by leveling the base, driving the pilings, mating the deck assembly, and installing well conductors and deck equipment.

The history of template platforms is impressive. In 1947 a steel platform was constructed in 20 ft of water offshore Louisiana. Later that year another platform of similar design was placed in 50 ft of water. One-hundred-foot depths were reached in 1955. By 1965, platforms were sitting in depths up to 225 ft, and by 1970 the record depth stood at 370 ft.

1976 saw a platform installed in 850 ft of water, and by 1978 the present record of 1,025 ft was set (Fig. 1–4).

While early platforms were supported by a veritable forest of thin pilings, later use of larger-diameter legs led to the reduction in number of support pilings. The number of wells operating on a single platform also has grown to as many as 96.

By 1982, there were over 3,400 fixed platforms operating throughout the world; over 98% of these were the pile-supported, steel-jacket template type. This figure, while imposing, makes manifest the difficulty in attempting

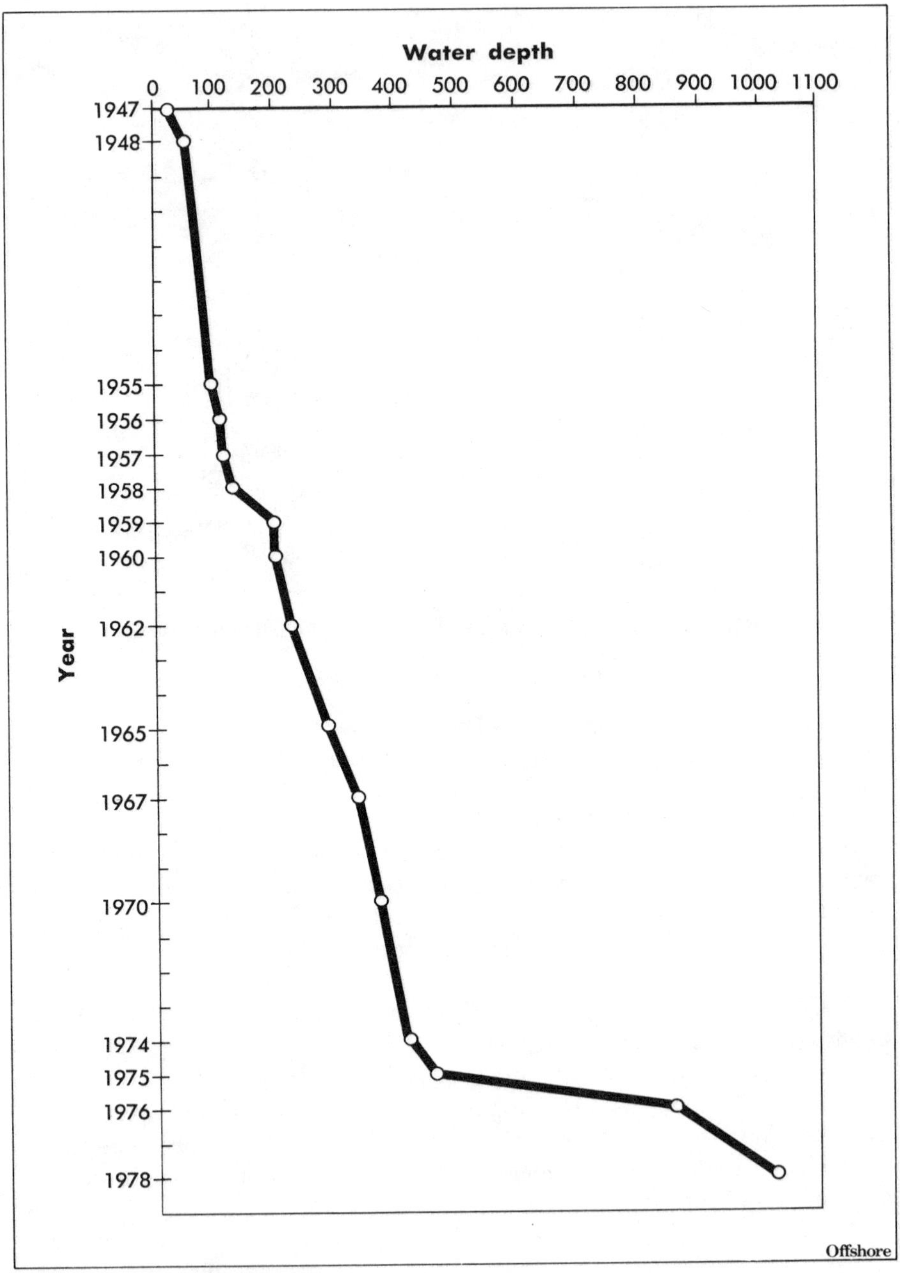

FIG. 1–4 Water depth records for platforms.

to set forth a detailed description of all of the various types of template platforms one may encounter in offshore waters. To present a more detailed portrait of a single template platform, it may be good to examine the largest of them all—the Shell *Cognac* platform (Fig. 1–5).

While certain design characteristics make the *Cognac* more atypical than other examples, the sheer magnificence of scope in its design and installation may give us some sense of how far platform technology has come and perhaps tell us what may lie ahead in offshore structure design.

FIG. 1–5 Shell's *Cognac* Platform.

DRILLING AND PRODUCING OFFSHORE

In 1976, Exxon installed what was up to then the deepest offshore structure in the world, the *Hondo* platform, in 850 ft of water off the coast of California. This platform was noteworthy for its overall height of 945 ft and because of its unique fabrication and assembly techniques. Due to the size of the jacket, which was too heavy for conventional handling, a sectionalized-jacket procedure was used. That is, the jacket was fabricated in one piece then separated into two sections that were launched and reassembled while floating in a horizontal position. The reassembly utilized locking hydraulic flanges and full-strength welds made from inside the legs.

Meanwhile, Shell's *Cognac* platform was under development. Because the structure was to be installed in 1,025 ft of water, it was again felt necessary to employ a sectionalized-jacket approach. However, in this case there were to be three sections, fabricated separately and installed individually by stacking, i.e., by first lowering the base section, leveling and driving the skirt piles, and only then lowering the other sections one at a time onto each other. Each section was connected to the next by inserting 72-in. diameter tubular connecting pins, 600 to 800 ft long, that would be grouted within the legs to provide extra strength and stability.

The base section itself weighed 14,000 tons and was anchored by twenty-four 84-in. diameter pilings driven almost 500 ft into the subsea floor by an underwater pile driver. This hammer, used for the first time on this project, was 40 ft long and 10 ft wide.

The middle section of the jacket was 315 ft long and weighed 8,000 tons, and the top section was 530 ft long and weighed 11,000 tons. These sections were in turn lowered by cable from derrick barges onto the base section and, with the addition of an eight-legged, 2,000-ton deck with two drilling rigs, the entire structure towered 1,099 ft from mud line to derrick top. A total of 59,000 tons of steel were used, including pilings and connecting pins. The *Cognac* had 60 straight and two curved well conductors. Installation of this massive structure was completed in 1978.

While the *Cognac* today represents the ultimate in deep-water template platforms, future developments in technology may bring even more deep-water capabilities, although restraints are more likely to be imposed by economics rather than by technical limitations. Meanwhile, template platforms will continue to dominate the offshore structure population. They are a known quantity of proven reliability that have an exemplary record of providing a safe, stable, and efficient drilling and production facility.

CONCRETE GRAVITY PLATFORMS

Although some very small concrete gravity platforms were built in the Gulf of Venezuela during the early years of platform development, not until the great surge of offshore activity in the North Sea during the 1970s was there serious consideration of concrete gravity drilling and production structures. Since the installation of the first large concrete gravity platform in the North Sea in 1973, their numbers have grown to nearly 20 worldwide. And although most of these structures are still located in the North Sea, others can now be found in areas as diverse as the waters off Brazil, the Congo, and Louisiana. New and exotic designs for concrete or steel gravity platforms are now under study for such sites as the shallow Arctic waters from Siberia to Alaska and for several deep-water locations off Canada.

A concrete gravity platform differs from the template platform in several dramatic aspects. Perhaps the most fundamental difference (apart from the use of construction material) lies in the method of anchoring the platform to the sea bed. Template platforms use pile-anchoring, that is, the legs of the jacket are fixed permanently to the sea floor by driving piles deep into the earth. Even the tension-leg and guyed-tower platform designs utilize some form of anchoring system, whether piles, anchored guy lines, or anchored tension cables. All of these systems require extensive, complicated, hazardous, and extremely costly underwater installation. A concrete gravity platform, on the other hand, sits firmly on the sea bottom and is held stably in place by the sheer force of its own broad-based massive weight.

Because of this unique approach, the most vital consideration in determining the feasibility of a concrete gravity platform for a specific site is the condition of the sea floor. In order to ensure platform stability, the sea bed must be virtually level, highly resistant to penetration, and extremely stable. For this reason, concrete gravity platforms are not appropriate for all offshore locations and applications. This inherent lack of adaptability, when compared with the great flexibility of applications and siting of template platforms, will no doubt ensure that concrete gravity platforms will remain a small, albeit valuable, segment of the offshore structure population.

Fabrication of concrete gravity platforms is usually a two-stage process, with early construction taking place in dry-dock-like, man-made basins. Later stages of construction take place in nearby protected waters, with deck mating completing the process before towing to the installation site (Fig. 1–6). The virtually 100% structurally complete platform is then ballasted, towed to sea, and finally lowered by controlled ballasting to its resting place on the sea

FIG. 1–6 Towing the concrete gravity platform to the installation site.

floor. This latter operation, while the most critical phase of the entire process, has been developed so expertly that even 300,000-ton structures have been lowered at controlled descent rates as low as 2 in./min to within 6 ft of the target site.

Following the late 1960s' discovery of the vast North Sea production potential, studies were undertaken to evaluate the feasibility of using concrete gravity structures for drilling and production. Sea-floor conditions and water depths were deemed appropriate, and several advantages of the new type of platform were postulated.

In addition to the fact that structural weight would provide the required stability, thus eliminating the need for pilings and the resultant long and expensive at-sea construction, other factors cited included the following:

a) Concrete is a relatively inexpensive material and is available in most parts of the world.
b) It is highly fire and explosion resistant.
c) Construction techniques are relatively simple and well known.
d) Transporting the structures is not a terribly complex task.
e) Concrete of the proper composition possesses high corrosion resistance.
f) Fatigue is not as great a factor as with steel under alternate stresses caused by the severe North Sea wave action.
g) Maintenance requirements are minimized.
h) Because of their hollow base designed for buoyancy during towing, the structures can provide enormous oil storage capacities.

Early doubts about the potential difficulties with immersion during installation were later resolved, as shown in the tremendous delicacy and accuracy achieved in final submersions.

Concrete gravity platforms are today found in waters as shallow as 12 ft in the Gulf of Mexico to waters over 500 ft deep in the North Sea. Total platform weights have reached astounding levels, as witnessed by the 600,000-ton *Ninian Central* platform or the even more massive 825,000-ton *Stratfjord B* platform in the North Sea. Operational depth and weight limits have not been strictly defined at present, although it is generally felt that the *Stratfjord B* will probably represent the upper limit on weight. Any design for deep-water sites would probably incorporate several innovative features, such as articulated anchoring devices, a high percentage of steel, and repositioning the ballast/storage chambers.

In some designs of gravity platforms, steel has already been extensively employed, primarily in the columns or legs and in the deck support structure, with concrete being used for the ballast and/or storage compartments and base sections.

Just as template platform designs have evolved into several distinctive types, so concrete gravity platforms have assumed various configurations and sizes. However, the basic design components have remained essentially the same (Fig. 1–7).

The bottom section of the platform is usually fitted with antiscouring devices that penetrate the sea floor to some degree. Depending upon the degree of evenness or settlement potential of the sea bed, either a continuous foundation slab or several foundation plates or caissons are used. Upon these

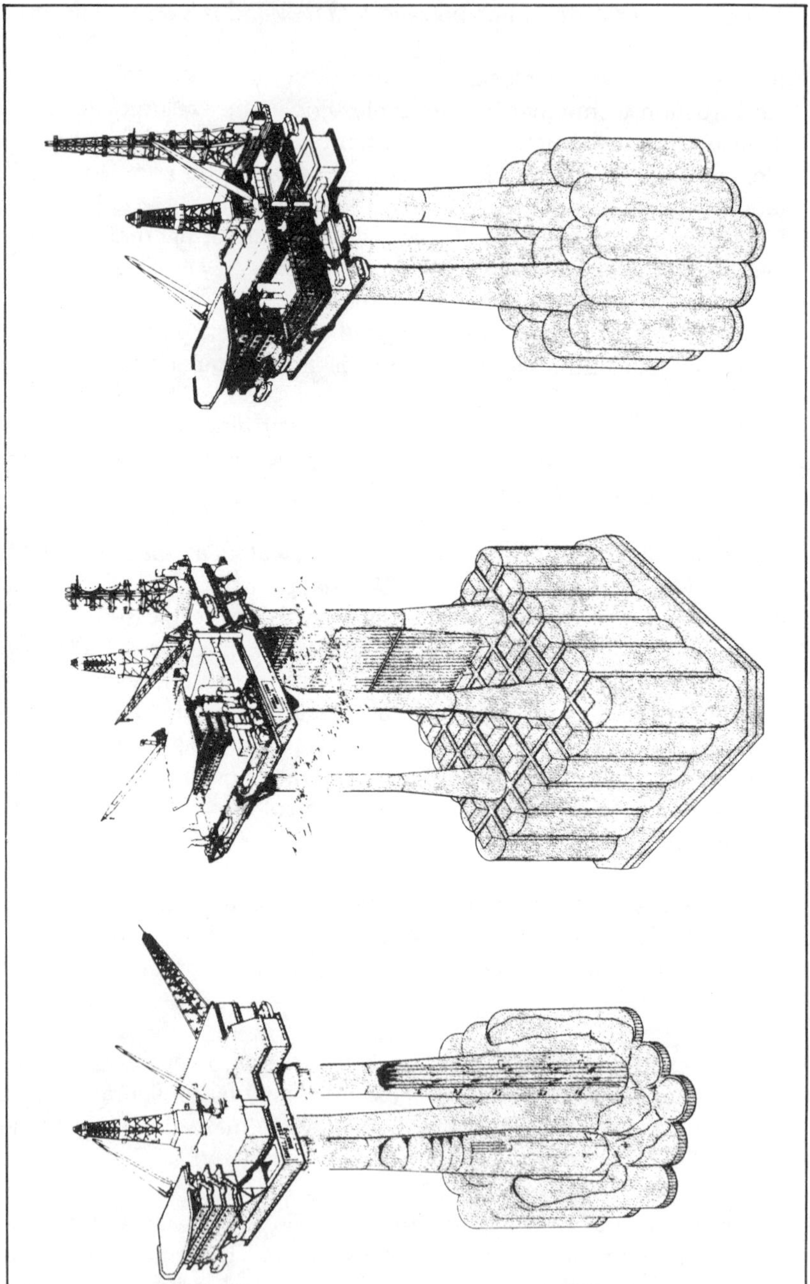

FIG. 1–7 Examples of types of concrete gravity platforms.

base sections several compartments or cells complete the bottom section of the platform.

These cells, along with compartments in the legs, perform a dual function. During towing, their ballasting provides the required buoyancy. Upon arrival on site, their controlled ballasting lowers the platform to the sea floor. During operations, the compartments or cells may assume their alternate role as oil-storage chambers.

Because the cells are not continuously filled to capacity with oil, sea water must be pumped in or out to provide the remaining ballast for the required stability. Storage capacity of the larger platforms can reach one million barrels of oil. The construction of these chambers can utilize either precast, prestressed concrete sections or more commonly the slip-form method of pouring.

Rising from the base section, the columns or legs reach past the water line to provide the necessary air gap between the deck and the water's surface. Whether a gravity platform will have one, two, three, or more concrete or steel support columns is based upon such factors as anticipated deck load capacity, total deck area, and additional capacity required for control systems, pumping equipment, or storage. The columns may also act as conductors for well installation, thus assuming some of the characteristics of a drill-through-the-leg platform.

Experience in the North Sea has shown that, for the same water depth, concrete gravity structures are more economical than the template platforms with estimated overall savings of up to 20%. However, not all designers or fabricators are in agreement that gravity structures provide a viable alternative to template structures. At the present state of technology, concrete platforms seem to have limited potential. And although during the late 1970s they were viewed by some as the wave of the future, the industry seems to have backed off from its original enthusiasm.

The market for new concrete or steel gravity platforms for the next several years appears to have shrunk, and several concrete yards in the North Sea area are standing idle. However, with continued advances in technology, new and more exotic designs, improvements in concrete composition, another upsurge in activity in the North Sea, or even the opening of new areas to intensive offshore drilling and production, we may once again see the demand for concrete gravity structures rise to impressive levels.

GUYED TOWER

With the tallest template platform in existence standing in 1,025 ft of water in the Gulf of Mexico, there are indications that—given the present state of technology—different approaches may be required for depths beyond the 1,000-ft range.

One new design concept that has been under serious study and development is the guyed-tower platform (Fig. 1–8). The guyed-tower platform differs significantly from a template platform. The tower itself, while bottom founded, is supported laterally by anchored guy lines attached to the structure near the water surface. These guy lines are attached to the deck in wedge-like clamps, pass vertically down the structure to fairleads approximately 50 ft below the water line, and thence extend radially at an approximate 30° slope to clump weights on the ocean floor. From the clump weights, the guy lines further extend to anchor piles. The number and size of the guy lines are dictated by several factors, including the size of the structure, water depth, degree of redundancy desired, and environmental conditions. The typical guyed-tower platform contains 16 to 24 wire-rope-type guy lines.

The guyed tower, along with the tension-leg platform, is a prime example of the compliant platform approach to the challenge of deep-water drilling and production. As deeper waters (1,000+ ft) are encountered, the dynamic interaction between waves and structure reaches critical limits for the template platform. For example, whereas a shallow-water (300 ft) template platform would have a period of less than 3 sec, the structural period for a 1,000-ft jacket may reach 4 to 6 sec, resulting in a considerable amplification of the energy transmitted to the structure by the operating sea state. Under these conditions, fatigue aspects may be critical and additional steel may be required to reinforce the structural joints and stiffen the members. This modification would be reflected dramatically in fabrication costs and installation difficulties. As a compliant structure, the guyed tower—rather than trying to minimize the dynamic wave/structure interaction by reducing the structural period below the period of the sea state—takes the opposite approach and lengthens the structural period to as high as 30 sec, in effect negating the amplification of dynamic load factors.

In October 1975 a one-fifth scale test model of a guyed-tower platform was installed in 300 ft of water in the Gulf of Mexico. After five years of testing, the concept was deemed both feasible and practical, and detailed design of a full-scale guyed-tower platform was begun. The installation of this first commercial drilling and production guyed-tower platform is scheduled

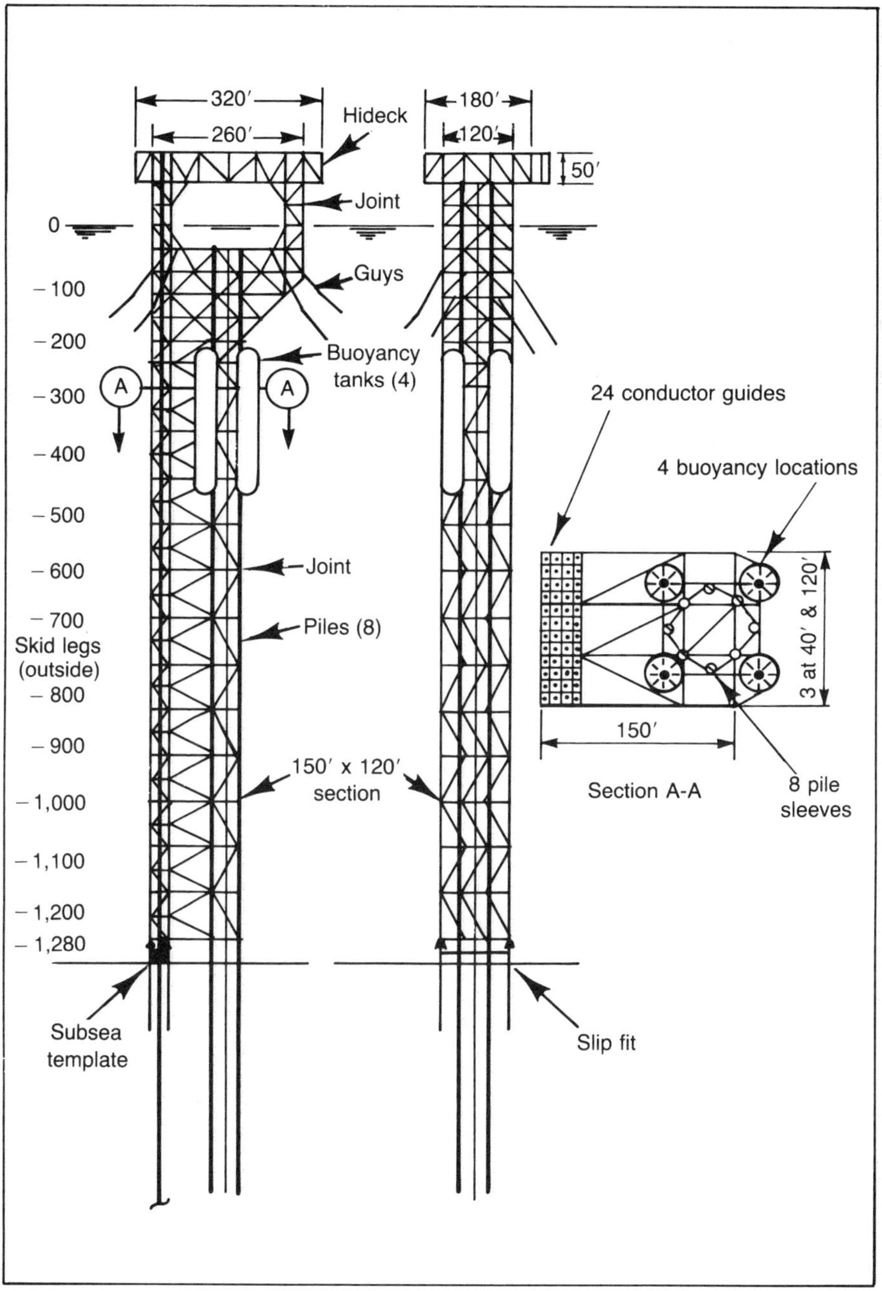

FIG. 1–8 Guyed-tower platform.

for late 1983 or early 1984 in the 1,000-ft water of the Gulf of Mexico off Grand Isle, Louisiana.

This tower, while generally incorporating the basic design successfully embodied in the test model, contains several significant modifications. The test tower uses a spud-can base under the theory that the weight of the tower will embed the can in the ocean floor. Originally, piles were not felt necessary because the foundation would primarily be resisting vertical loads. However, later studies indicated the spud-can approach could develop potential difficulties since there are too many unknowns relative to flexibility and settlement. As a result, the full-scale tower will be pile founded. The piles on the first tower will be driven 450 to 500 ft into the sea floor. This pile-founded concept for the guyed tower is now deemed most appropriate for 1000 + foot waters, although in shallower waters such as those encountered in the North Sea, the spud-can base will likely still prove more feasible.

Buoyancy tanks have also been added to the tower to assist in carrying part of the deck load. In the absence of a spud can, a 58-slot drilling template has been added whose piles will also extend upward into the tower's legs. Additional torque-resistant piles at the mud line will provide still more anti-fatigue capability. The deck configuration of the new tower has undergone modification from a modular concept to that of an integral design. The entire structure, including mooring system, will weigh 43,000 tons.

Studies have determined that the most economical range of operation for a guyed-tower platform falls in the 800- to 1,600-ft spectrum, with 2,000-ft depths constituting their probable limit because of fabrication difficulties.

Estimated overall costs of the guyed tower compare favorably with template platforms in waters where the two structures are compatible. Overall steel cost, construction time, and installation time provide a cost advantage to the guyed tower, especially in 1,000 + ft water depths. Although the total project cost of the first commercial guyed-tower platform (including $3.5 million for the test model) grew from a projected $500 million to a figure approaching $1 billion, future guyed towers are expected to reflect significantly less expense.

The performance of the first full-scale guyed-tower platform will be monitored closely. Although future guyed towers may incorporate significant modifications, the expectation is that the guyed-tower design will prove an economic and practical mainstay of drilling and production in 1,000- to 2,000-ft waters.

TENSION-LEG PLATFORMS

Sometime in late 1983, the world's first full-scale tension-leg platform (TLP), designed by Conoco, is scheduled to be installed in 485 ft of water in the North Sea. The introduction of this new design to the growing family of offshore platforms was preceded by 15 years of research and development work, including the at-sea testing of a one-third scale prototype off California in 1975.

The tension-leg platform is a buoyant structure held in place by vertical members anchored to the sea bottom. Submerged flotation cylinders designed to minimze the platform's response to weather and wave conditions provide the desired buoyancy. And since the platform's buoyancy exceeds its weight, the anchoring system is always in tension. The tension in the tethers limits heave, roll, and pitch while acting to reduce surge, sway, and yaw. And since the structure itself is not rigidly attached to the sea floor, any motion produced by earthquakes would be almost totally dampened by the time it reached the platform.

The TLP design concept includes the capability of ready adaptation to most specialized types of offshore structures. In addition to development and production drilling, the TLP can be used for other operations such as exploratory drilling, workover rigs, floating plant stations, and subsea well servicing. Unlike other fixed platforms, some TLP designs incorporate the capability of mobility. Thus TLPs are not limited to one location for the life of the structure and can be moved—for example, if delineation wells indicated the necessity for relocation. The TLP could also be moved from a depleted reserve to a productive field.

Just as in the case of floating drilling rigs, field development time for the TLP is significantly lower than for that of other platforms since it is possible to fabricate a production TLP before actual water depths and other design criteria are established. Given known environmental conditions and desired payload, hull construction can begin. As water depths increase, template platform fabrication and installation costs increase significantly. A major cost advantage of the TLP is reflected in increases associated with water depth, since on the TLP they are limited essentially to lengthening the anchoring members. However, with the TLP installation costs may be very high because of the complexity of hooking up the mooring system, template, riser, and wellhead assembly. The general consensus is that the most economically feasible range of operations of the TLP would be in water depths of 1,500 to 2,000 ft.

The original tension-leg platform prototype that was tested successfully during a five-month period in 1975 off the California coast had a triangular deck, three-leg design. It was anchored by cables in 200 ft of water. Later designs included a square hull, four-leg, one-template type and a hexagonal, six-leg, six-template type that would incorporate the through-the-leg drilling principle (Fig. 1–9).

The first full-scale tension-leg platform to be built and installed will be located in the Hutton field in the North Sea and will have a rectangular hull, six legs, and one template. Detailed design and fabrication of the $1.3-billion project began in 1981 with simultaneous installation of a 32-slot sea-bed template in 485 ft of water. The original eight-leg design for this platform was modified to facilitate barge-assisted mating of the hull and deck assemblies. Another modification was an increase in the number of 10½-in. mooring tubulars from 12 to 16, four in each corner leg. This modification was prompted by the anticipated increase in probable deck load.

The hull and deck will be fabricated separately on land, launched, and then mated just offshore. After mating, the now-integrated, single-piece unit will be towed to the well site.

The 243 × 260 ft deck is only 37 ft high, making its center of gravity very low and enhancing the platform's stability. Unlike template platforms where the deck is merely a support frame for facilities and equipment, the TLP deck is an integral part of the structure, in effect holding the hull together.

Overall platform stability of the TLP, when compared with the template and guyed-tower types, showed that vertical movement was virtually negligible, thus comparing favorably. Under normal sea conditions, surge, sway, and yaw will be small; but under extreme 100-year storm conditions, lateral displacement could reach 75 ft, although up to 50 ft of horizontal movement is not debilitating.

The first tension-leg platform is expected to be the forerunner in the growing utilization of new nonconventional deep-water platform designs. Although no present plans are envisioned for TLPs in the Gulf of Mexico, a $6 million study is underway for a TLP in the 1,000-ft waters offshore California. By late 1983, the eyes of the offshore industry will be sharply focused on the new Hutton field TLP. Whether it performs in a productive, economical, and safe manner will no doubt significantly influence the future development and construction of tension-leg platforms.

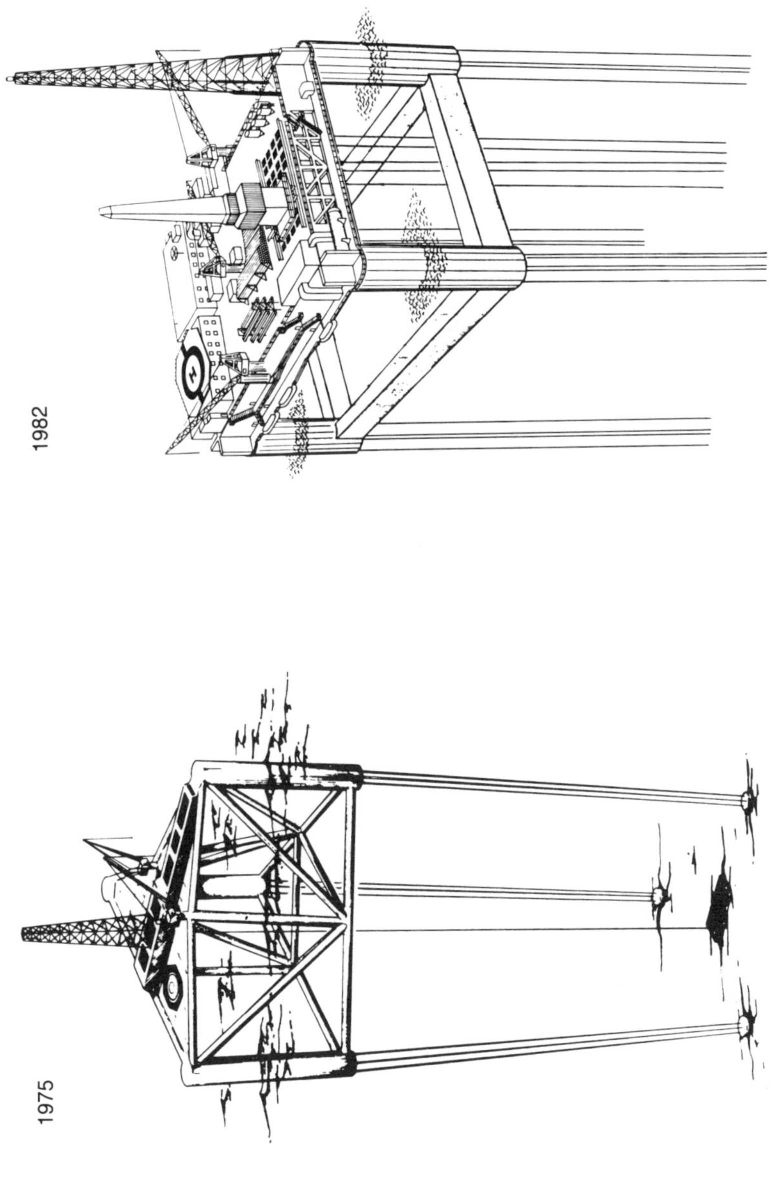

FIG. 1–9 Prototype and actual tension-leg platforms.

DRILL-THROUGH-THE-LEG PLATFORMS

In 1969, Hurricane Camille struck the Louisiana Gulf Coast area. Three template platforms that had been constructed to withstand hurricane winds and waves were overturned or suffered extensive damage. After an investigation, it was determined that an even more powerful force than wind and waves had come into play: that of underwater mudslides. Investigators found that mudslides could more than triple the loads of an offshore structure.

As a result of this discovery, a crash program was undertaken by Chevron USA to develop a type of platform to withstand the mudslide threat. These efforts eventually led to a reconsideration of an existing design known as a drill-through-the-leg platform (DTL).

By the late 1960s, a number of DTL platforms had been installed in the Cook Inlet of Alaska. These shallow-water structures had been designed to combat the effects of sea ice in the Arctic waters.

Unlike template platforms, where conductor pipes must be provided for each well in order to protect the surface casing against natural forces, the legs of these platforms—containing several well slots—would act as one large conductor. These early platforms had proven effective in providing protection, stability, and strength and were felt to offer substantial savings in platform structural weight and, as a result, lower costs.

The drill-through-the-leg platform had already been under consideration for use in the Gulf of Mexico based upon the probable reduction in costs over template platforms. But in the wake of the mudslide problems typified in the hurricane-prone waters seaward of the Mississippi River mouth, it was decided that the same protective characteristics shown by the DTL platforms in Alaska could be used to advantage in the Gulf.

The conventional drill-through-the-leg platform design that had been developed for Gulf coastal waters does not appear radically different from a steel-jacket template platform. The DTL design has four steel legs; two are vertical and the other two are smaller in diameter and are sloped or battered. By use of deep plate girders, the deck load is transmitted to the piles. Drilling is done through the two large vertical legs, while the battered legs are used for additional stability and, in some cases, fresh-water storage. From 1971 to 1982, eleven of these conventional DTL platforms were installed in the Gulf of Mexico in waters ranging from 145 to 264 ft.

In order to combat the unusually large loads confronted in the mudslide areas, the conventional DTL design was modified by making the jacket legs

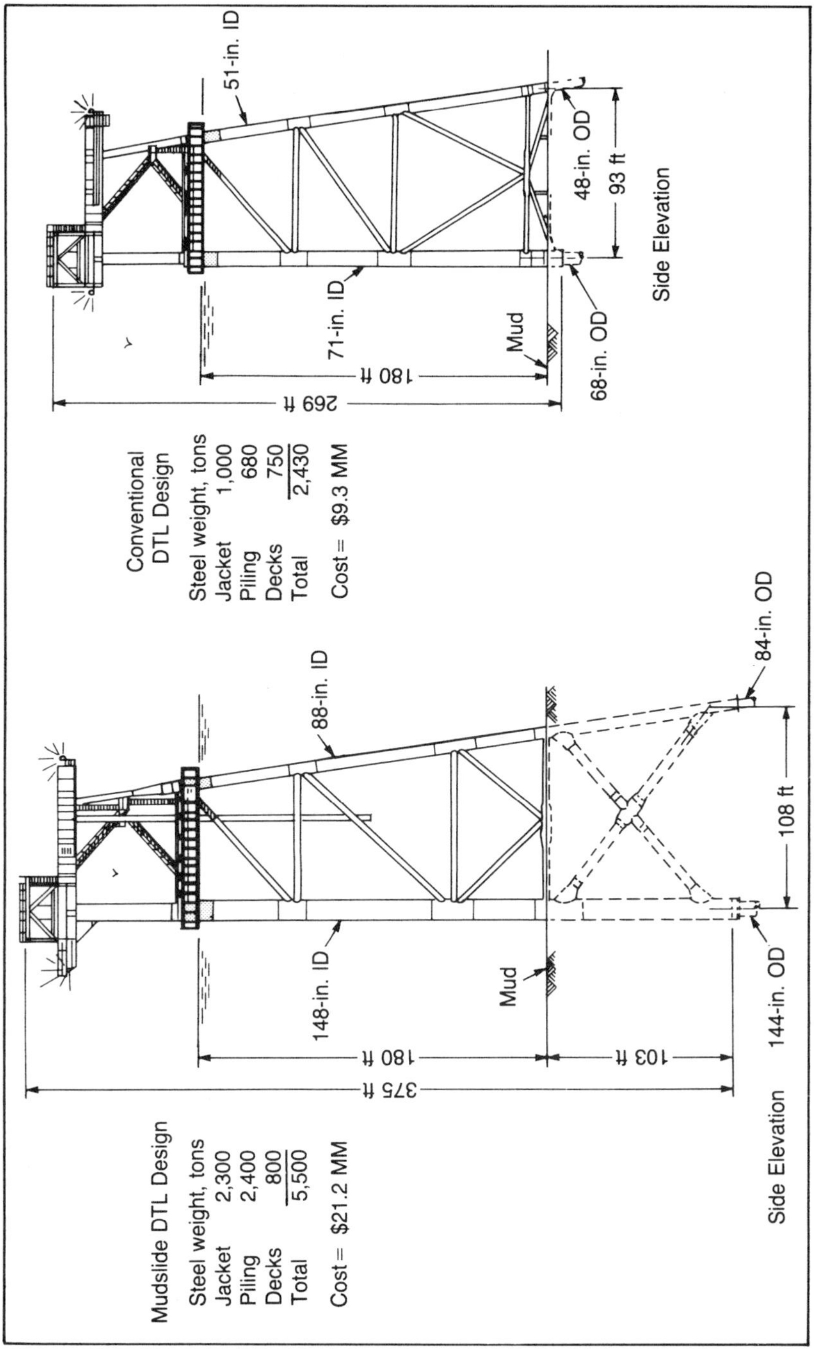

FIG. 1–10 Drill-through-the-leg platforms.

and pilings much longer, larger, and thicker (Fig. 1–10). Because of the soil conditions in the mudslide area, the jacket was extended 100 ft or more into the sea floor below the mud line where pilings would be driven even deeper. The modifications in size were not insubstantial. The drilling legs were enlarged from the conventional design's 48-in. diameter with three well slots to massive 148-in. diameter, 24 well slot legs. Structural weight was increased dramatically, and fabrication and installation times were greatly extended. As a result, overall costs of these mudslide platforms increased to two or three times the costs of platforms in other parts of the Gulf of Mexico. By 1982, only four of these modified DTL mudslide platforms had been built.

One major limitation on the eventual size and effectiveness of drill-through-the-leg platforms is the size of the components required for completion, e.g., connectors and controls. A more detailed discussion of this problem is found in Chapter 2.

Even though there may be physical limitations to future developments and the increased costs associated with mudslide-type DTLs may make those types economically unfeasible, the conventional drill-through-the-leg platform has proven to be an efficient and cost-effective drilling and production structure. Its effectiveness in providing well protection under special environmental conditions will always make it an attractive alternative to template platforms when certain unique offshore situations are encountered.

Table 1–1 Evolution of Platforms

	Total platforms in existence	Water depths (max. ft)	Platform weight (max. tons)
Conventional			
1950	40	75	NA
1960	308	198	1,650
1970	928	376	7,700
1975	2,308	474	44,000
1980	3,149	1,025[d]	59,000[d]
1982	3,594	1,025	59,000
planned 1983-85	4,000+	1,800[e]	NA
Concrete Gravity			
1975	3	462	384,500
1980	17	551[f]	601,220
1982	19	551	825,000[i]
planned 1983-85	22	551	825,000
Guyed Tower			
1975	1[a]	300	3,000
planned 1983-85	1[b]	1,020	43,000
Tension Leg			
1975	1[c]	200	698
planned 1983-85	2	485[g]	46,900[g]
Drill-Through-the-Leg			
1970	3	130	4,000
1975	6	145	5,500
1980	15	264[h]	8,400[j]
1982	18	264	8,400
planned 1983-85	NA	NA	NA

[a]Exxon guyed-tower test model, offshore Louisiana, 1975-79
[b]Exxon guyed-tower, installation scheduled late Petrole 1983 off Grand Isle, La.
[c]Deep Oil Technology X-1 test model, offshore California, 1975
[d]Shell *Cognac* platform, offshore Louisiana, 1978
[e]Steel-jacket platform under study for Chevron, offshore Spain
[f]Shell *Cormorant* platform, North Sea, 1977
[g]Chevron Hutton field, North Sea, scheduled for 1984
[h]Conoco DTL, offshore Louisiana, 1980
[i]Mobil *Stratfjord B,* North Sea, 1982
[j]Chevron *D* DTL, offshore Louisiana, 1980

Sources: *Offshore, Oil & Gas Journal, Ocean Construction Locator, Ocean Industry,* Institut Francais du Petrole

Table 1–2 Production

	1960	1970	1975	1980
World yearly offshore crude oil production (millions of barrels)	77.1	1,711.2	1,990.0	2,237.4
% of total world production	7	16	17	21
U.S. yearly offshore crude oil production (millions of barrels)	47.3	337.1	303.1	395.8
Average production per platform (barrels per day)	685	5,347	2,362	1,947

Sources: *Offshore,* National Supply Co., Institute Francais du Petrole

In 1960 there were only 23 countries where offshore exploration activity had taken place. By 1980 this figure had grown to over 100 countries. During the same period countries producing offshore oil had increased from only seven in 1960 to 35 in 1980. Offshore production grew from an 11% share of the world's total oil production in 1960 to a 21% share (17% for natural gas) in 1980. The 1980 total of producing offshore wells in the world was more than 20,000, located in approximately 800 producing offshore fields. Over 75% of the world's offshore platforms are located in the Gulf of Mexico. Virtually all concrete gravity platforms are located in the North Sea.

Table 1–3 Total Number of Platforms by Region

	1981	Projected by 1985
United States	2680	2840
Latin America (including Carribean)	171	279
North Sea	99	155
Mediterranean	56	63
West Africa	80	104
Middle East	179	250
Southeast Asia (including India)	181	257
Far East and Australia	17	23
Total	3463	3971

In the early years of platform development, virtually all platforms were located in the Gulf of Mexico. By 1981, 77% of the world's platforms were still located within U.S. waters. Because of the upsurge in planned construction in other countries, this percentage may drop to approximately 71% by 1985. Areas such as Latin America, the North Sea, and Southeast Asia are expected to experience a rapid growth in their offshore platform population during the 1980s. However, even with this contemplated new construction, the overall growth rate for worldwide platform installations is not expected to approach the incredible boom of the 1970s when the worldwide platform count more than tripled.

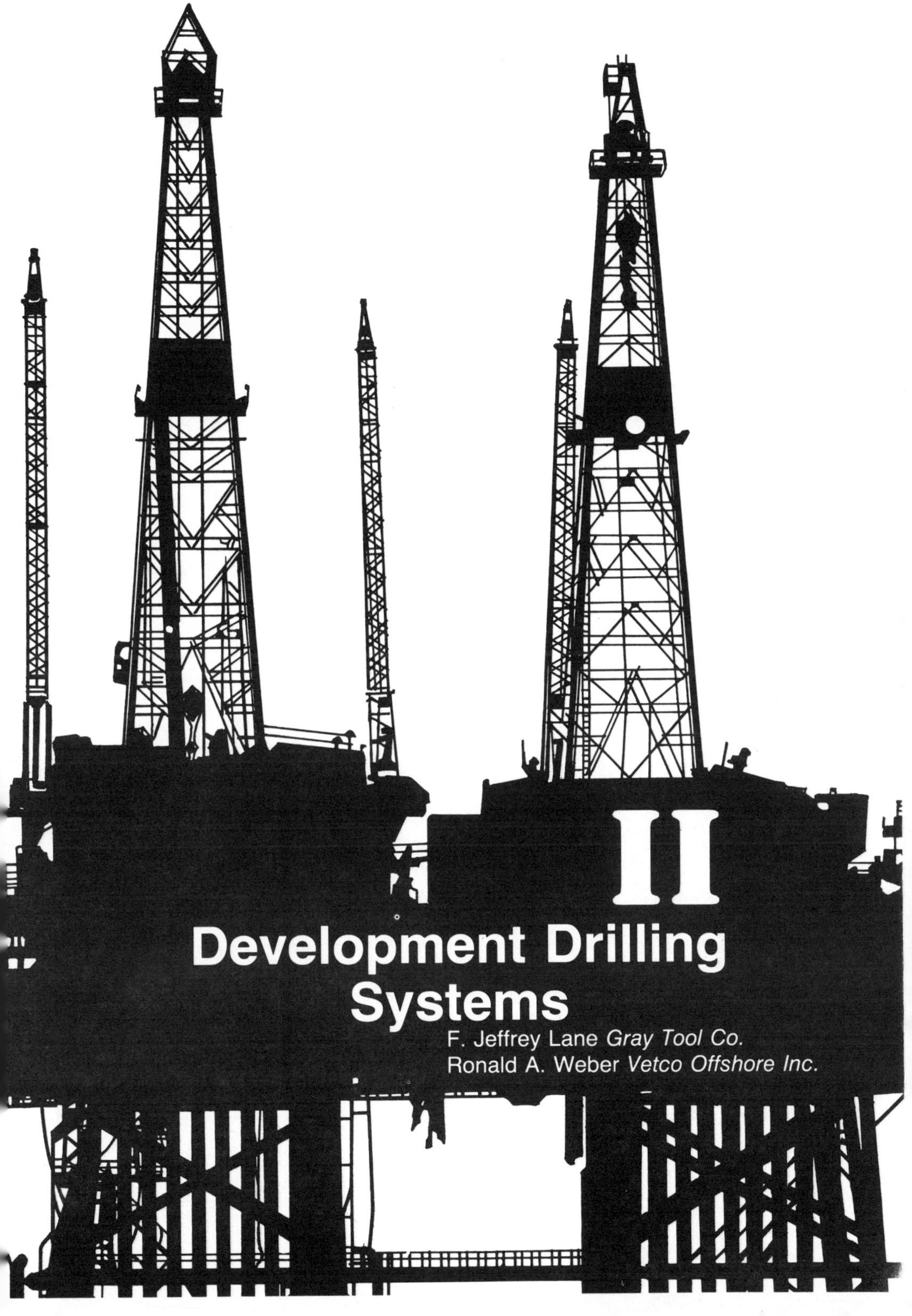

II
Development Drilling Systems

F. Jeffrey Lane *Gray Tool Co.*
Ronald A. Weber *Vetco Offshore Inc.*

2

Through-the-Leg Drilling

Through-the-leg wells are drilled and completed inside a structural leg of an offshore platform. From single wells to ganged multiples within the same leg, this technique offers an efficient means of protecting the well from special environmental conditions and is a potential savings in platform fabrication costs.

PROTECTING THE WELLS

By installing the wells through the platform legs, the legs can be utilized to protect the surface casing while providing stability and strength sufficient to resist external forces such as sea-floor mudslides, a glacier-like action found in certain offshore locations such as the Mississippi Canyon off Louisiana. Ice movement is another environmental condition that may justify through-the-leg wells. This type of solution to sea ice problems is exemplified, in the extreme, by the monopod platforms found in Alaska's Cook Inlet, where all wells are contained in the platform's huge single leg (Fig. 2–1).

In a conventional platform, individual conductor pipes must be provided for each well to protect the surface casing from the natural forces of the sea as well as geomechanical conditions such as mudslides (Fig. 2–2). Along the U.S. Gulf Coast, these conductors are typically 26 to 36 in. in diameter. With the through-the-leg method, these conductor pipes are eliminated and replaced in function by the platform leg. Installing several wells within the same leg can multiply the advantages inherent in this savings.

Installing the wells inside the platform legs also reduces the wetted area of the platform exposed to sea and wave action by reducing the hydrodynamic

FIG. 2–1 Monopod platforms, such as this installation in Alaska's Cook Inlet, are extreme examples of through-the-leg drilling and completion. *(Courtesy Gray Tool Co.)*

FIG. 2–2 Conductor pipes extending from conventional platform completions greatly increase the hydrodynamic profile of the platform structure. *(Courtesy Gray Tool Co.)*

profile of the platform structure through eliminating individual conductor pipes. This may allow further savings in platform structural weight and, hence, cost.

PACKAGING THE SYSTEM

Obviously, there are limitations on how far this practice can be carried. There are limits to the size of the completion equipment that may be utilized and the number of wells that can be installed practically in a given leg. Structural problems and complexity multiply with increasing leg diameter, and additional reinforcement of the leg may be required. Also, installation considerations for the platform itself must be considered in determining maximum feasible leg diameter.

A major limitation in packaging through-the-leg systems is the physical size of completion components and assemblies. Although a few through-the-leg programs have been completed using flanged connections of components within the leg and in the completion systems, the concept has been most successful where special, proven piping connectors facilitate reliable makeup of joints in restricted space; simplified access to service, replace, and maintain well-system components; and additional flexibility in installation and hookup. For these reasons, most through-the-leg installations utilize the Grayloc® connector system (Fig. 2–3).

Use of these connections, designed to API safety criteria, has permitted installation of up to 12 wells in a 120-in.-diameter platform leg (Fig. 2–4). Using flanged connections, a similar program required a leg diameter of 150 in., resulting in a significant increase in platform weight and cost, estimated to be an additional $2.5 million.

A factor that will become more important as petroleum production in deeper and more hostile waters increases is the thermal differential between the produced well fluids and surrounding sea water. The relationship between the internal and external components of a well string is exaggerated by wider temperature differentials. As the length of the string exposed to differential temperatures becomes greater in deeper water, so the effects of expansion *vs* contraction of the well components accumulate. The improved isolation of the actual well string from the surrounding water with through-leg wells helps inhibit this differential stress.

FIG. 2–3 The Grayloc connector permits easy assembly of well components in a much smaller space with a great deal less weight than flange connections. *(Courtesy Gray Tool Co.)*

FIG. 2–4 A 12-well installation in a single 120-in. diameter platform leg was made possible by using the Grayloc clamp connector. *(Courtesy Gray Tool Co.)*

By concentrating the wells in a smaller space, made possible by eliminating the large-diameter individual well conductors, the overall size of the platform deck space required can be reduced with a potential savings in weight and cost (Fig. 2–5). The concentration of well bores into a small area also results in reduced horizontal movement required to set up the drilling rig for drilling each well. Thus, a rig with a smaller slot size may be used with potential savings in rig cost, and setup time per well is reduced.

FIG. 2–5 Through-the-leg wells concentrate components in a small area, potentially permitting smaller platform deck size. *(Courtesy Gray Tool Co.)*

SYSTEMS DESIGN

Essential to successful application of the through-the-leg multiple well technique is implementing the systems design philosophy. This concept relates all elements of the project to a unified whole (the system) with a hierarchy of subsystems. Because of the concentration of so many elements and operations into so small a space, complete integration of all of the parts into the whole must be considered from the outset and extended through the life of the reservoir.

At the point when field development planners begin consideration of a platform, systems design should begin. Ideally, consultation between those responsible for field development and platform design should include the wellhead equipment manufacturer; drilling company; service, maintenance, and installation interests; platform fabricator; and reservoir engineers. This mix should represent a full spectrum of field-wise talent able to consider the entire project.

Before design can begin in earnest, a number of basic questions must be answered. How many strings of casing will be run? How many wellheads will be installed per leg and of what size? What size tubing and what size valves are required? What pressure ratings of equipment will be necessary? How many tubing strings are there per well? Will injection wells be required? How about enhanced recovery pumps? What kind of automation system is required for surface valves and downhole control valves? Once these questions are answered, the design elements can begin to be culminated and the relationship of the well cluster to other topside production components can be developed.

While a typical platform requires little documentation in the form of plans, specifications, and procedures for the actual well equipment, it is likely that considerable attention will have to be paid to equipment layout and design for a through-the-leg program. System designers must anticipate possible later changes in tubing and equipment combinations and configurations, recovery methods, well programs, and field substitutions for components. As system complexity increases and space allotments for each well decrease with the increase in the number of wells per leg, the requirement for thorough documentation becomes apparent.

Although through-the-leg well systems are a relatively recent development, the internal (and many external) components are the same as those used in more traditional drilling and completion. Sometimes the method and sequence of installation is different. Tighter installation tolerances may also require techniques slightly different from standard oil-field practice. Orientation of well components becomes critical in high-density situations, and this may also require additional documentation (Fig. 2–6).

A comprehensive job of design and documentation can often find and cure problems before the system reaches the field. With through-the-leg systems, this is extremely important, as the generally tighter working space around neighboring wells will usually make the normal oil-field "custom fitting" techniques less effective.

Future maintenance, replacement, and workover may also be expedited by referring to design/installation documentation and avoiding unnecessary disassembly or incorrect disassembly/assembly sequences.

FIG. 2–6 Orientation of well control components becomes critical in high-density clusters of wells. *(Courtesy Gray Tool Co.)*

It is important that specifiers and designers of through-the-leg systems maintain a sense of standardization. This will allow a maximum transfer of field experience in the installation, operation, and maintenance of the system while controlling development costs and reducing delivery time. There is also the additional confidence gained by using well-proven component parts and subsystems. As in Gray Tool Company's proven through-the-leg installations where special equipment configurations are necessary in the interest of access, existing internal assemblies with long records of reliable oil-field service are utilized.

In the same manner, many specialized components that have been generated for through-the-leg systems have found application in other situations. Examples of this type of specialized or custom equipment are tubing hangers with relocated control-line ports and dual valve bodies reconfigured to let all controls reach outboard around the circumference of the platform leg yet retain all standard internal components.

ACCESS AND FLEXIBILITY

Since through-the-leg wells normally are configured using as much standard, off-the-shelf hardware as possible, the emphasis is on the practical, compact packaging of this equipment. In fact, a real challenge for the designer is to retain adequate access to all well components during installation, hookup, and test and then to retain access for later service, replacement, and workover. Complicating this task may be future requirements for enhanced recovery methods using one of the various forms of injection, downhole pumps, or other techniques.

Because the future cannot be predicted accurately, eventual recovery methods are hard to forecast and real-world conditions encountered after the project is installed in the offshore oil field are virtually impossible to foresee. Field necessity may someday require substituting a valve, packoff, or replacement part. For these reasons, flexibility must be designed into the system from the start, realizing that space and accessbility limitations are inherent.

Access to all necessary areas of the well equipment must be planned completely, from setting the first surface casing through the drilling and completion of all wells to the final hookup. Simultaneously, the inevitable service, maintenance, and replacement of the well and its component parts must be considered.

Experience in this relatively new technique can save both time and money while improving the sophistication and flexibility of the system. Great strides have been made in a relatively short span of time as the benefits of through-the-leg development have become apparent.

INSTALLATION

The proximity of one well to another and the limited space allocated for each well's equipment package dictate both a requirement for care in guidance and alignment of well strings during installation and the maintenance of relatively close (API normal) tolerances during installation and assembly. The key to a proper installation is getting off to a good start. Since the surface casing will pilot the rest of the well string, its correct positioning and alignment is critical to the orientation of subsequently installed equipment. A template or drill jig is generally provided by the through-the-leg system manufacturer to help drill the wells in the proper locations (Fig. 2–7). This template and conscientious attention to tolerance control ensure precise alignment and guidance with relative ease.

With the platform leg serving as conductor and the surface casing attached to the drilling template at the top of the leg, installing the balance of the casing program can then proceed in a normal fashion. In fact, with the exception of the conductor pipe, through-the-leg wells utilize the same casing programs as any other platform well. Consequently, casing hardware is standard, off-the-shelf equipment.

Current designs for surface casing guidance and alignment systems fall into two categories: those that position the casing at the top of the platform leg only, and those that provide additional alignment and guidance of the strings throughout the length of the leg. Both types are designed to carry the weight of the surface casing strings during installation and until the strings are cemented in place. In the case of the latter system, the alignment and guidance components within the leg may also serve as additional structural supports for the leg and the casing.

Using the technique by which a template provides only positioning and spacing at the top of the leg, a typical installation would consist of the following operations. The open top of the leg is cut off and carefully leveled. Then a blind cap with a central flange is installed to guide the hole-opener bit while drilling out the full-diameter pile hole inside the leg and below (Fig. 2–8). The blind flange is then removed, the internal pin pile is driven (if required), the leg top is dressed and releveled, and a template with holes for each well's surface casing is welded to the top of the leg (Fig. 2–9). The template must be carefully oriented and leveled before welding since it will serve as the primary guide for the entire multiwell cluster installation. The well index holes in the template are slightly oversized to allow final close-tolerance alignment later and to permit clearance for passage of the casing joints.

FIG. 2–7 A well bore template is usually supplied by the wellhead manufacturer to ensure alignment and guidance of surface casing will be held within tolerance. *(Courtesy Gray Tool Co.)*

All surface casings will be run and cemented before any drilling takes place. A moveable slip bowl and slips are provided for each well's surface casing. These hang and position the strings until final adjustment and secure the casings to the template (Fig. 2–10). Normally, casings for all wells in the leg will be lowered to a given level, hung in the slips, and then lowered again to a deeper level. This procedure helps keep the strings of casing relatively parallel as they are run.

FIG. 2–8

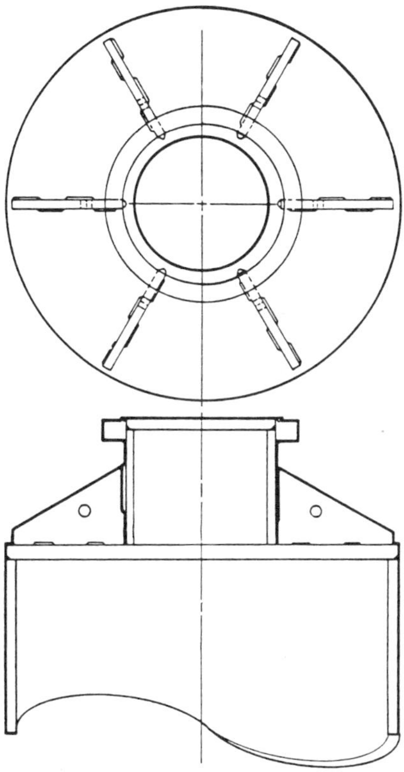

FIG. 2–9

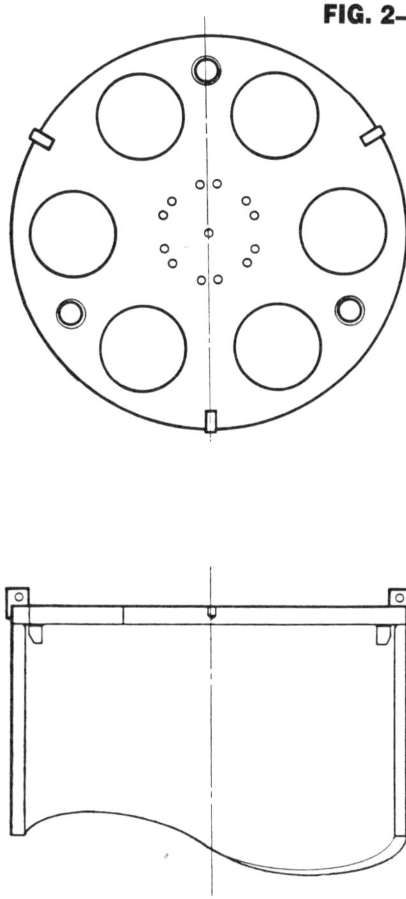

FIG. 2–8 Temporary blind platform leg cap with central flange guides the hole-opener bit. *(Courtesy Gray Tool Co.)*
FIG. 2–9 Leg-top template insures correctly positioning and indexing well bores and surface casing. *(Courtesy Gray Tool Co.)*

Each string should be made up with a mud line suspension housing at the appropriate level before the entire string is run. The depth at which the suspension system is to be located must be determined by the operator as part of the well program. It is very important to maintain an accurate record of the depth to the mud line hanger housing so that subsequent hangers can be installed easily.

FIG. 2–10

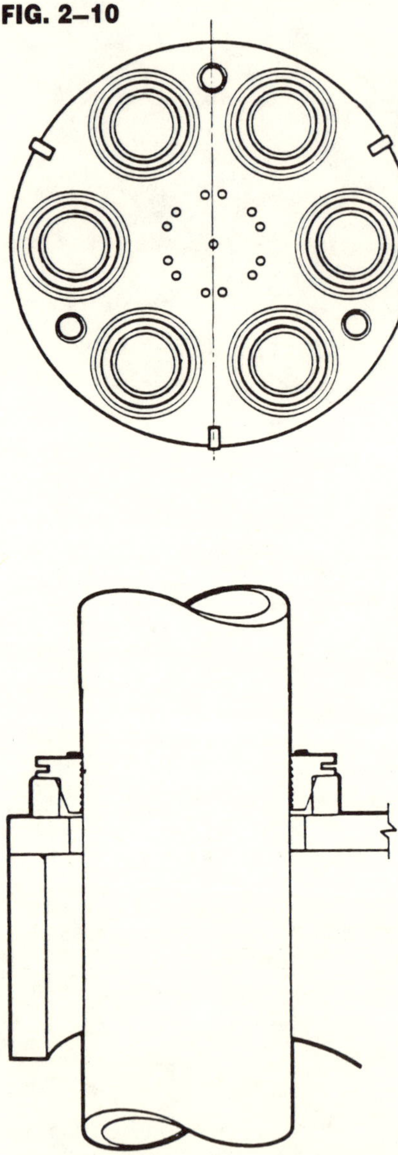

FIG. 2–11

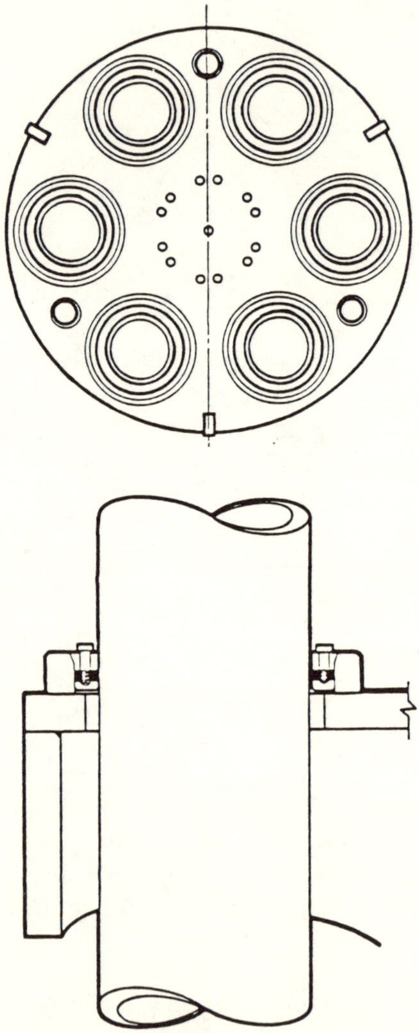

FIG. 2–10 Surface casing is hung temporarily in template by slips in movable slip bowls. *(Courtesy Gray Tool Co.)*

FIG. 2–11 After casing is cemented, slips are replaced with packoffs seated in the slip bowls. *(Courtesy Gray Tool Co.)*

After all surface casing has been run to the required depth, it is cemented normally through grout holes in the template and is cut off at a specified distance above the template surface. Each casing must now be oriented and attached permanently to the template. After the cement has set, slips are removed and replaced by packoffs seated in the slip bowls (Fig. 2–11). A tool or jig for aligning and positioning the casing is provided by the wellhead system manufacturer to locate these points accurately without the need for making critical measurements on the rig.

Using this alignment tool, each casing is positioned correctly and the slip bowls are then welded solidly to the template (Fig. 2–12). Casing heads normally are installed and tested, and equipment installation for drilling the first well can commence.

When additional guidance and alignment of the casing strings is required, a system that provides one or more alignment points within the platform leg, in addition to the fixture at the top of the leg, may be utilized. These alignment points may be built into the leg while the platform is being fabricated in the yard or may be installed in the field along with the casing.

Installation begins by assembling the combination cementing and separator head (the bottommost aligning device) to the lower separator and stabilizer. The casing separators within the leg also reinforce the surface casing structurally and act as stabilizers. The assembly, which is installed on a central pipe string made up of drill pipe or casing, is then lowered into the platform leg until the lower separator and stabilizer can be hung off on a strongback placed across the top of the leg (Fig. 2–13).

An appropriate number of spacer spools are added to the central pipe string supporting the separators, and the assembly is lowered further into the leg. This process is repeated until all separator and stabilizer assemblies and required spacer spools are in place within the leg (Fig. 2–14). A base plate is then affixed to the top of the string and lowered to rest on the top of the platform leg. The base plate is positioned, oriented, and welded to the leg (Fig. 2–15).

At this point, the drilling rig can be brought into position and the cementing string made up to the cementing sub. The cementing sub is then lowered on the cementing string through the inside of the central shaft until it reaches the bottom of the guidance assembly, where it can be threaded into the cementing and separator head. The cementing and separator head is then disengaged (J-slot) from the lower separator and stabilizer and is suspended

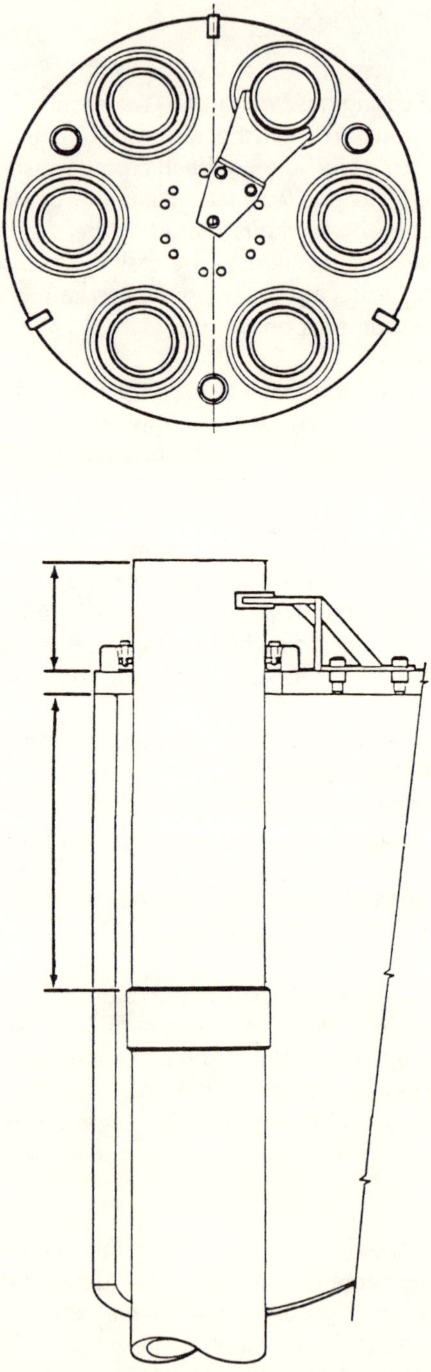

FIG. 2–12 Casing is oriented by using the fixture supplied by the wellhead manufacturer, and slip bowls are welded solidly to the template. *(Courtesy Gray Tool Co.)*

FIG. 2–13

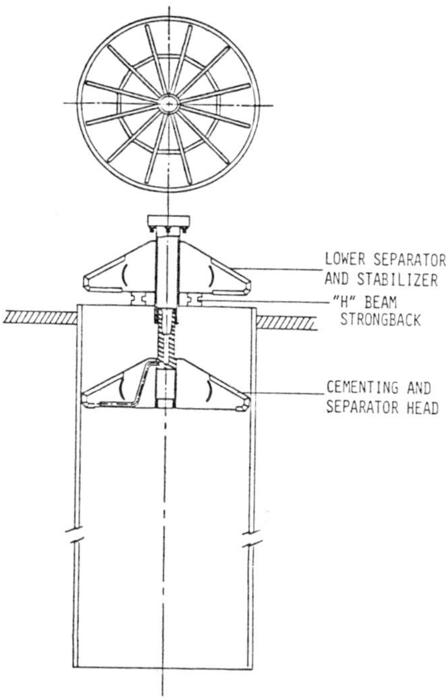

FIG. 2–14

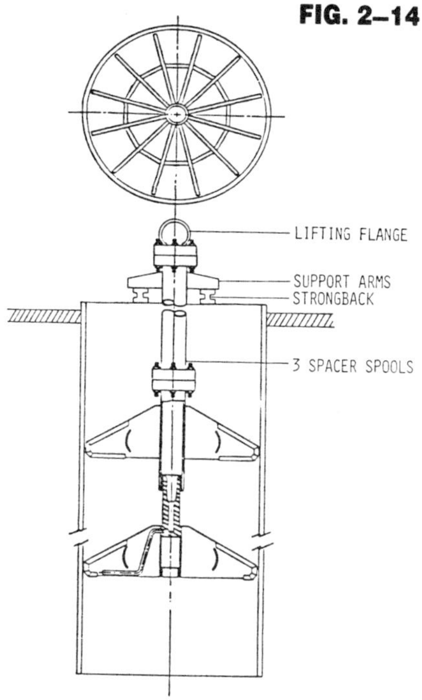

FIG. 2–13 The lower separator and stabilizer assembled to the cementing head are hung off the top of the platform. *(Courtesy Gray Tool Co.)*
FIG. 2–14 Spacer spools with an additional separator and stabilizer assembly are run into the leg. *(Courtesy Gray Tool Co.)*

from slips at the base plate. Surface casings may then be run in to a depth that allows them to hang below the cementing and separator head.

As each string is run, it is suspended by slips at the base plate. When all of the surface casings have been run in to this point, each string is lowered individually to a point below the cementing and separator head, but slightly above the bottom of the open hole, and hung off (Fig. 2–16). After all strings are hung off, they are trimmed off at a specified height above the base plate. The slips supporting the cementing string are then removed, and the cementing and separator head is lowered to hole bottom.

As the cementing and separator head is lowered, it provides a combing effect to align the casing strings the entire depth of the hole. The cementing

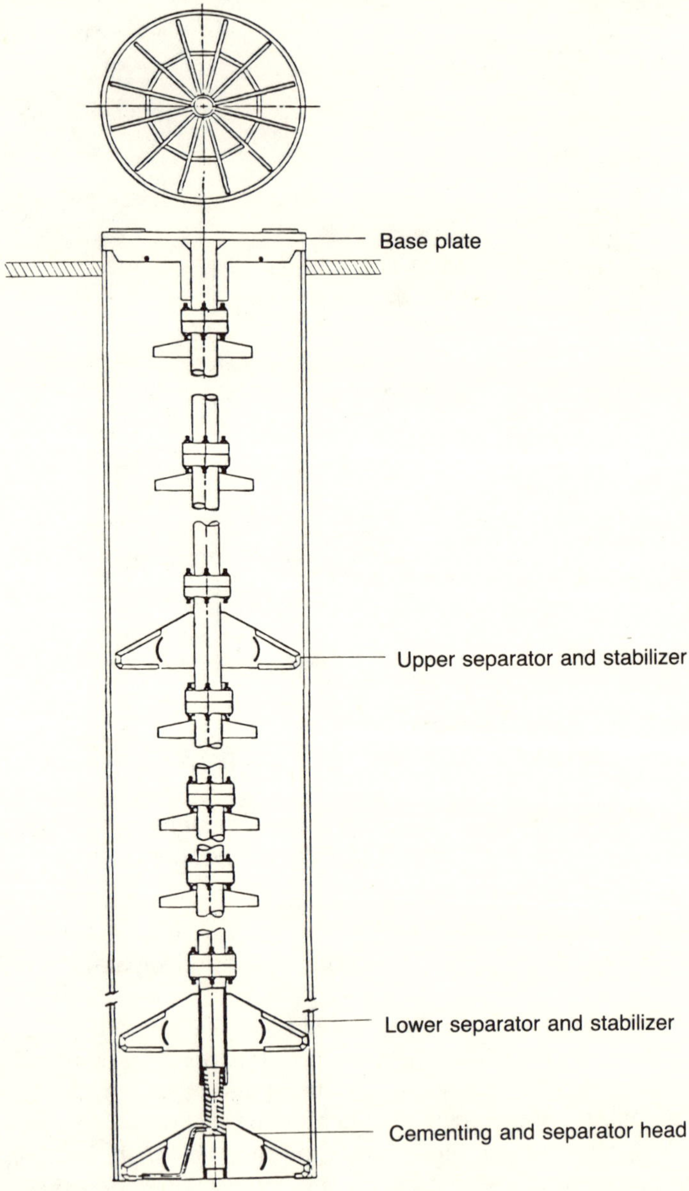

FIG. 2–15 When all required separator-and-stabilizer assemblies are in place within the leg, the base plate is welded to the top. *(Courtesy Gray Tool Co.)*

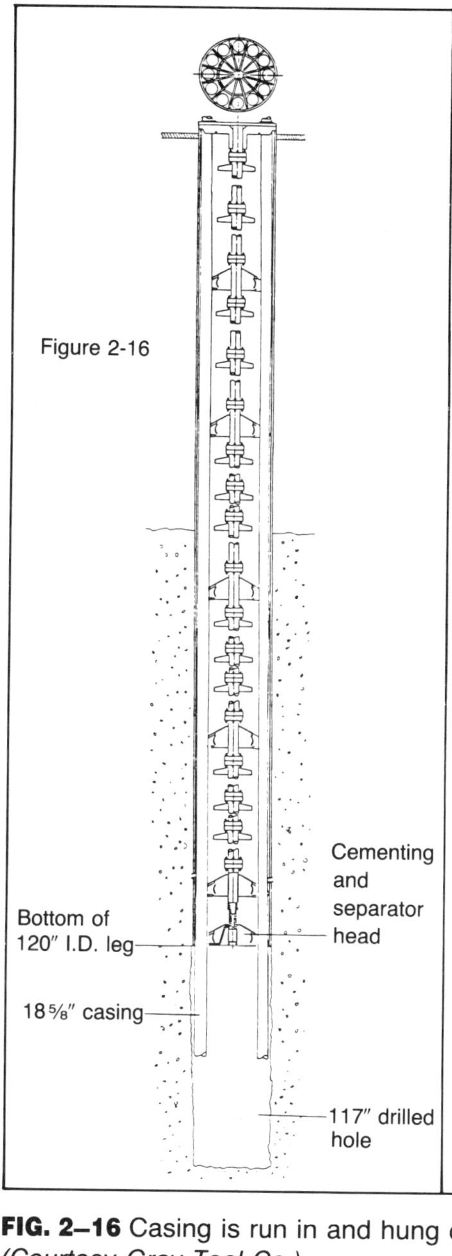

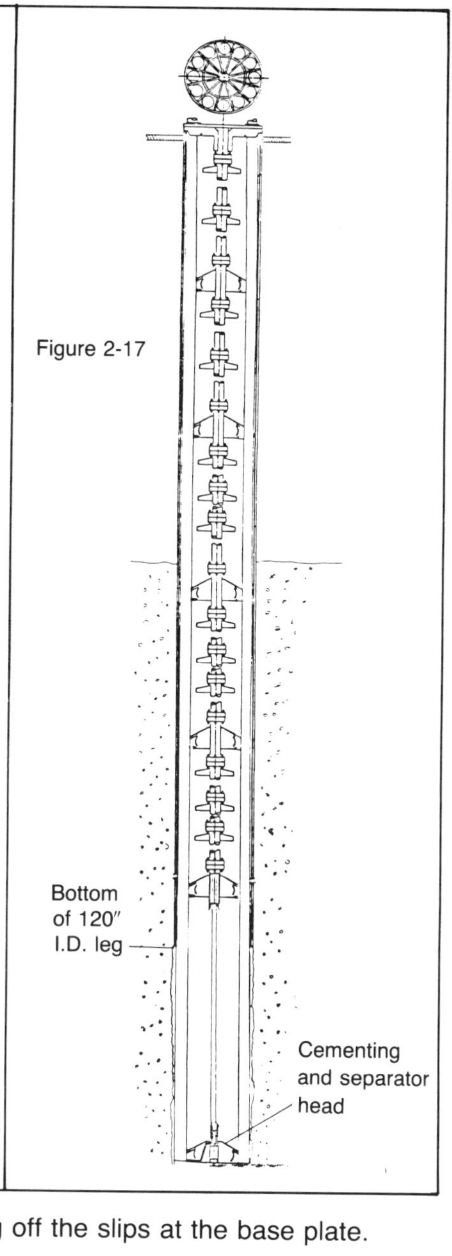

FIG. 2–16 Casing is run in and hung off the slips at the base plate. *(Courtesy Gray Tool Co.)*
FIG. 2–17 The cementing head is lowered to the hole bottom, "combing" the casing string; cementing is accomplished. *(Courtesy Gray Tool Co.)*

and separator head is left on bottom and cement is run through the cementing string and head to fill to the desired level in the platform leg (Fig. 2–17). After the cement has set, the temporary slips hanging the casing strings from the template are removed and replaced by split packoffs. Drilling equipment can then be attached and drilling can begin.

With either of these methods, tolerances and alignment of the wells can be held with only reasonable attention. This is possible because the system provided by the wellhead manufacturer has been designed to eliminate most field measurement and judgment except for what could be considered normal oil-field practice.

Cementing procedure is an application of standard methods. Where modification of normal practice is required, special equipment is provided as part of the through-the-leg system package. Good systems design always considers the installation, which in this case must include the real-world conditions of offshore drilling using standard rigs and equipment. When planning through-the-leg multiwell installations, a complete system is required—not merely hardware.

DRILLING

The actual drilling operations in through-the-leg wells are the same as with any other platform well. The proximity of other wells requires extra attention to drilling control, and the need to maintain working access to completed wells also requires special consideration.

Through-the-leg wells can also provide advantages in terms of early production for improved cash flow. As an example, several wells in a leg can be completed and placed into production while the drilling rig moves to a well cluster in another leg. When the wells in the second leg are drilled and completed, they can be placed into production, allowing those completed wells in the first leg to be shut in while the balance of the wells in the first leg cluster are drilled and completed. This practice of alternating leg drilling provides early product flow, and continuous flow may be maintained while development drilling is carried on at a safe distance from producing wells.

Another time- and money-saving technique involves the use of two rigs: a normal rig for drilling and a lighter unit for completion work. By allowing the completion rig to work while drilling proceeds at another leg well cluster, elapsed time from platform emplacement until full production level is reached can be reduced. A cost savings may also be realized due to a reduced on-site time requirement for the heavier drilling rig with its higher day rates.

Drilling in the same leg where there are completed wells can be accomplished safely, but it requires care in drilling control because of the proximity. Maximum distance between drilling and completed holes is desirable (such as drilling on the opposite side of the leg from a completed well), and good practice requires the nearby wells be shut in while drilling is in progress.

The close spacing between wells in a leg cluster and the need to maintain access to all sides of the well control and drilling safety equipment usually will require that the blowout preventer (BOP) stack be elevated above the height of the installed production equipment. Drilling operations carried out in the same leg where there are completed wells require a drilling stub extension that raises the BOP stack above the top of the entire Christmas tree.

SUMMARY

Through-the-leg drilling can offer many advantages, particularly where environmental factors such as soil or ice movement pose problems. The technique may also allow cost savings in platform design and construction in some instances and may allow earlier product flow than conventional field development.

Although the multiwell cluster within a platform leg is a somewhat new configuration, most of the well equipment is standard and the drilling and completion work is simply tailored operations of standard oil-field practice—modified slightly to fit the circumstances.

Basic to successful utilization of the through-the-leg well concept is total integration of the components and operations into a designed system. The wellhead manufacturer normally is responsible for system design, and choice of this source of engineering, design, and equipment is a critical early decision in through-the-leg projects.

3

Floating Drilling

Floating drilling vessels are classified in two basic categories: (1) ship shape, e.g., drillships with hulls similar to ships and barges that are self-propelled or towed to location (Fig. 3–1) and (2) semisubmersibles (semis) (Fig. 3–2). Each type has advantages and disadvantages that help determine which type vessel is best suited to a particular application. Generally speaking, drillships are selected for their rapid mobility, lower day rates, larger storage capacity, and deep-water capability. The semi may be preferred because of its stability, particularly for drilling in rough seas. Barges with drilling rigs are used normally only in sheltered areas but offer economy of operation.

Drillships date all the way back to 1953 when an ex-naval patrol craft, the *Submarex*, was converted to drill core sample holes to a depth of 3,000 ft off the coast of California. Three years later, the drillbarge *Cuss I* became the first full-scale floating drilling vessel. In 1961, the *Cuss I* was fitted with four manually controlled swiveling thruster units for dynamic positioning (DP), allowing the vessel to drill without a mooring. The first custom-built DP drillship, Global Marine's *Challenger*, became fully operational in 1968.

Station-keeping ability is important in determining the amount of time a drilling vessel can operate effectively. In recent years, requirements for the ability of DP drilling vessels to maintain a given position have increased dramatically, enabling them to challenge deeper waters and more hostile environments—thus enhancing oil and gas recovery opportunities. Many of today's dynamically positioned drillships, utilizing computer-controlled thrusters, are drilling in water over 2,000 ft deep, and some are drilling in waters of 5,000 to 10,000 ft (Fig. 3–3).

Drillships with conventional anchoring systems usually drill in less than 2,000 ft of water. A typical drillship anchoring system consists of 8 to 12 anchors, radially spaced from the bow and the stern of the vessel. Another

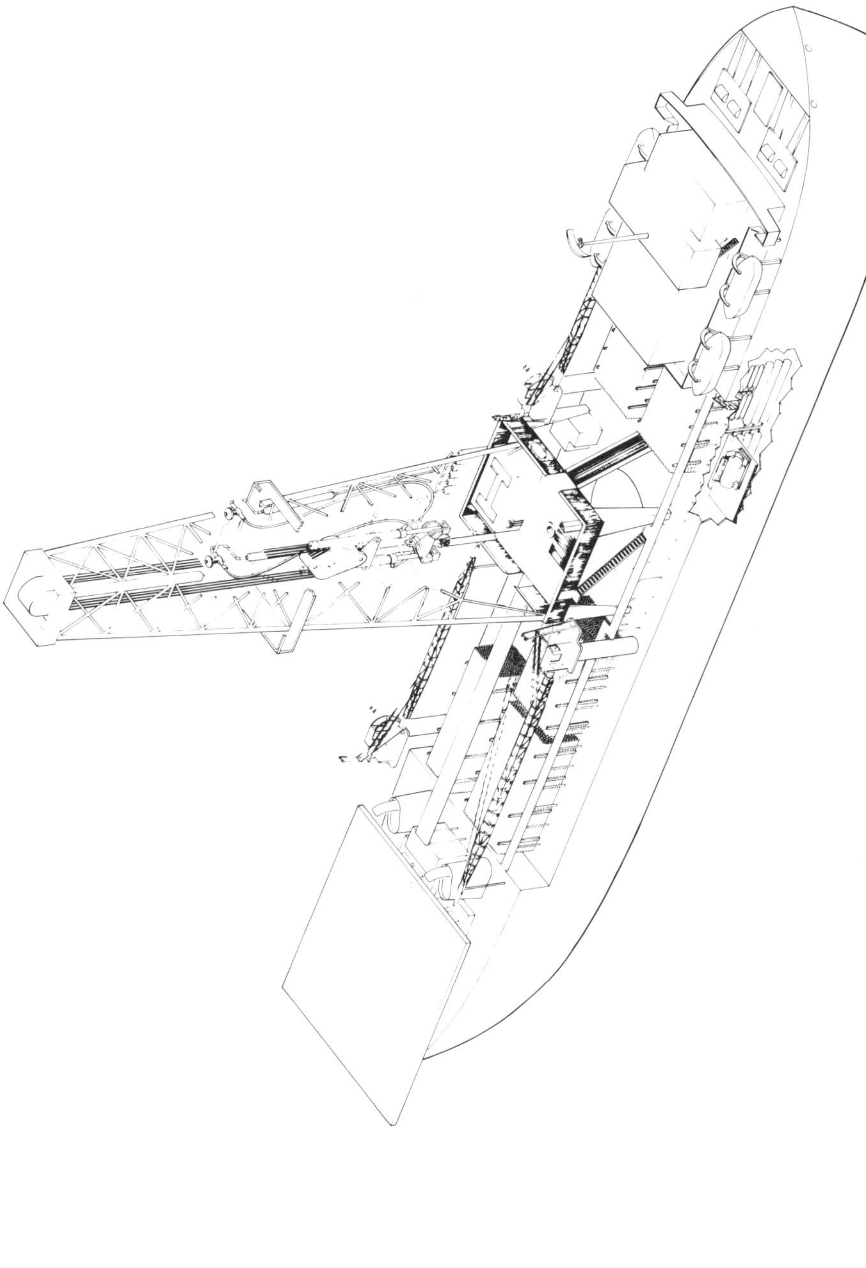

FIG. 3–1 Dynamically positioned drillship. *(Courtesy Vetco)*

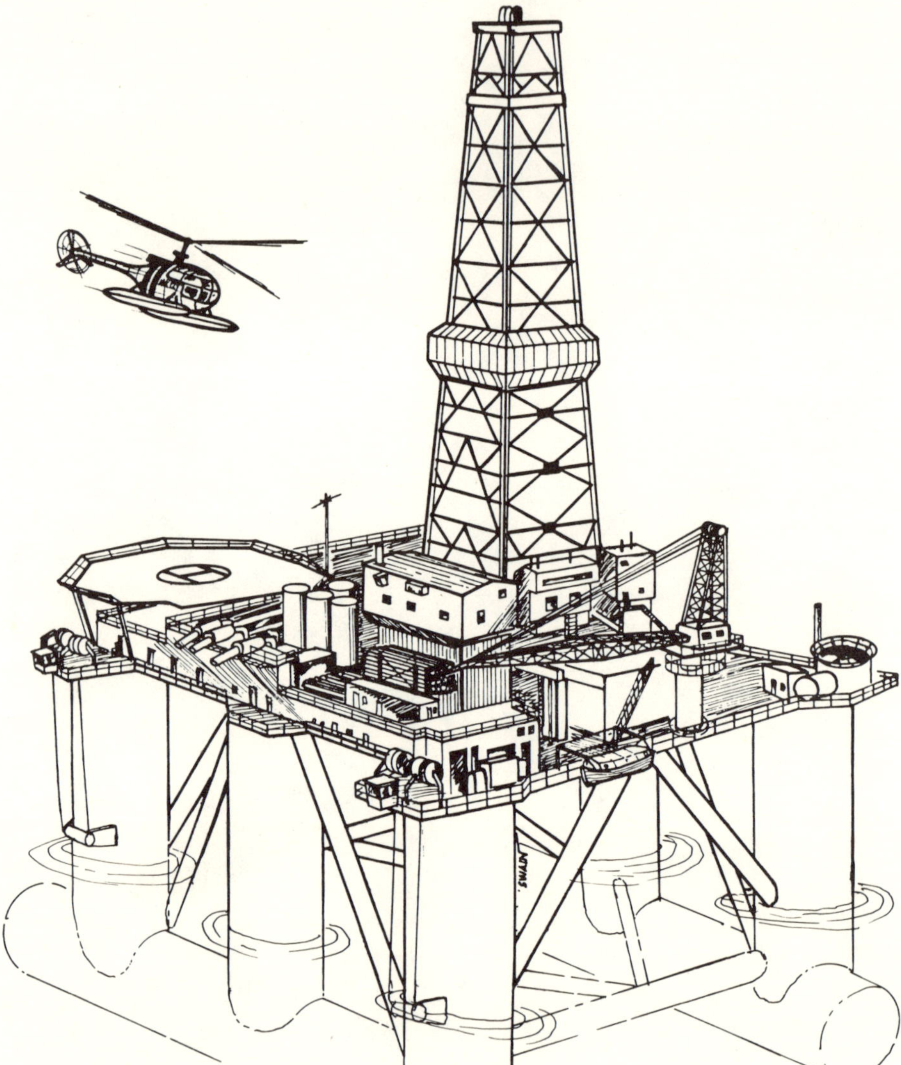

FIG. 3–2 Dynamically positioned semisubmersible. *(Courtesy Vetco)*

FIG. 3–3 A modern drillship.

system, turret mooring, allows 360° rotation of the drillship in relation to the mooring spread, thereby affording the ability to maintain the ship's heading into the weather. Drillships use subsea blowout preventers connected to the subsea wellhead housing.

Semisubmersibles have many deck configurations. The basic shapes are triangular, rectangular, and pentagonal. The first triangular semi, *Ocean Driller,* appeared in 1963, followed by *Bluewater II,* the first rectangular semi, in 1964. These vessels, supported by legs (usually three to ten) from deck structure to submersible hulls, generally drill in water depths from 200 to 2,000 ft. In 1970, semis were equipped with thrusters to facilitate towing and positioning over the well (Fig. 3–4). Most semis built after 1973 have some form of self-propulsion.

FIG. 3–4 Semi anchored and drilling. *(Courtesy Vetco)*

The first DP semi, the *Sedco 709*, was commissioned in 1978. Dynamically positioned semis still are not common, generally due to the power requirements for positioning the massive structures. However, the advanced-technology DP system offered in one new ice-class semi design claims station-keeping with enough precision to achieve three times the downhole drill rate of other hostile-environment semis.

In addition to station-keeping, other factors considered when selecting a drilling vessel are motion-compensation ability, load and stowage capacities, location of the rotary table relative to the vessel's center of gravity, mobility, and the experience and skill of the personnel. The primary advantage of the semi over a drillship in rough seas is its stability—its resistance to capsizing.

Since the semi's major buoyancy members are located below the surface, wave action on the vessel is diminished. Not only is heave greatly reduced, but the natural roll period is much longer for the semi. In spite of these favorable motion characteristics, a semi is subject to more limited loading conditions since its deck is 40 to 70 ft above the water line during drilling. Drillships can store tubulars and bulk materials low in the hull, while the semi must stow them on deck. (Weight located high on a vessel decreases stability; adding weight below increases stability. When loaded, then, the high deck of the semi increases the possibility of capsizing.)

Where rapid mobility is critical, drillships have an edge over semis. In calm seas, a ship's speed will be from 8 to 14 knots. Most of the fully loaded self-propelled semis travel at less than 8 knots. For this reason, a drillship usually is preferable when drilling only one or two wells in a remote area. However, downtime due to vessel motion must also be considered.

Wellheads and Casing

In many ways, a subsea wellhead performs much like a land wellhead—it supports the blowout preventer stack, suspends and supports the casing strings during cementing, and seals off the annuli between the casing strings while drilling or production operations are conducted. However, there are some important differences between floating drilling and land or platform operations. For instance, the following applies only to floating drilling:

- *The first and second casing strings are cemented with cement returns to the ocean floor.*
- *The last joint of casing is made up to a casing hanger so it may be suspended permanently prior to cementing.*
- *Flowby ports provide for mud returns.*
- *After a casing string is landed, cement slurry is pumped down the casing and casing shoe and up through the casing/hole annulus.*
- *Cementing plugs are released remotely.*
- *Casing seals must be remotely run and set.*
- *Special tools are required for testing.*
- *Wear bushings must be used to protect the wellhead.*

Pipe diameters may vary, depending on conditions and anticipated depth, but typically follow the standardized sizes, grades, and pressures (Table 3–1).

The hole sizes and casing programs used in offshore exploration drilling are essentially the same whether a jackup drilling vessel (using a mud line

Table 3—1

GENERAL BIT, CASING, WELLHEAD, BOP STACK AND RISER SIZES USED IN OFFSHORE DRILLING					
HOLE/BIT SIZE	**36″**	**26″**	**17½″**	**12¼″**	**8½″**
GENERAL CASING SIZES	30″ x 1.000″ Wall	20″ x 0.438″ Wall	13⅜″ 68 lbs/ft	9⅝″ 47 lbs/ft	7″ 23-35 lbs/ft
GRADE OF STEEL	A25 or 5L GRADE B	x52 Line Pipe	J55 - N80	C75 - N80	N80 - P110
AVERAGE SETTING DEPTH RANGE BELOW MUDLINE	150′ - 400′	500′ - 1500′ +	2000′ - 4500′+	5000′ - 12,000′+	6000′ - 20,000′+
METHOD OF SUSPENSION	From 30″ housing in Permanent Guide Structure	From either 16 ¾″ - 20 ¾″ 18 ¾″ or 21 ¼″ housings landed in 30″ housing	Run and landed with 13⅝″ profile housing in 20¾″ two stack system or from casing hangers inside of 16¾″, 18¾″ or 21¼″ housings using single stack systems.	From inside 13⅝″ housing in two stack system or inside 16¾″ - 18¾″ or 21¼″ housing using single stack system.	From inside 13⅝″ housing in two stack system or inside 16¾″ - 18¾″ 21¼″ housing in single stack system. Also can be suspended in 9⅝″ casing on liner hanger.
WELLHEAD AND BOP STACK SIZE & NORMAL PRESSURE RATINGS	30″ Conductor housing: BOP not normally connected to this housing	Single Stack System a. 16¾″ 5 & 10,000 psi W.P. b. 18¾″ 10,000 psi W.P. c. 21¼″ 10,000 psi W.P. Two Stack System d. 21¼″ & 20¾″ 2 & 3000 psi W.P. e. 18¾″ - 15 & 11″ - 15,000 psi	e. 13⅝″ 5 & 10,000 psi W.P.		
MINIMUM BORE THROUGH WELLHEAD HOUSINGS	26.750″	16¾″ - 5 = 15 11/32″ 16¾″ - 10 = 15 11/32″ 18¾″ - 10 = 17 9/16″ 21¼″ - 10 = 18⅝″ 20¾″ - 2 = 18¾″ 21¼″ - 2 = 18¾″ 20¾″ - 3 = 18¾″	13⅝″ - 5 = 12⅜″ 13⅝″ - 10 = 12⅜″ 13⅝″ - 15 = 12⅜″ 11″ - 15 = 9⅜″		
RISER SIZES (USUALLY GRADE X—52 SEAMLESS PIPE IS USED)		a. 18⅝″ & 20″ x 0.500″ Wall b. 20¼″, 20½″, 21″ & 22″ x 0.500″ Wall c. 24″ x 0.500″ Wall d. 24″ x 0.500″ Wall	e. 16″ × 0.500″ Wall f. 16″ × 1.000 Wall		

suspension system) or a floating drilling vessel (using a subsea wellhead system) is drilling the holes. The setting depths and grades of casing are in no way standardized (Fig. 3-5). The outline is only a generalization of what is practiced most frequently.

When drilling a hole for a particular casing string, the size of the hole obviously must exceed the size of the casing. Other factors that influence the hole size are shale formations that may expand into the hole and reduce the original hole size, joint ODs that are usually larger than the OD of the casing unless flush joint casing is used, and sufficient casing-to-hole annulus required for adequate circulation and cementing.

In some areas, regulatory agencies dictate the number and size of casing strings in a particular drilling application. However, abnormal formation pressures also may necessitate use of up to six casing strings. Corresponding equipment modifications have led to the development of many specialized

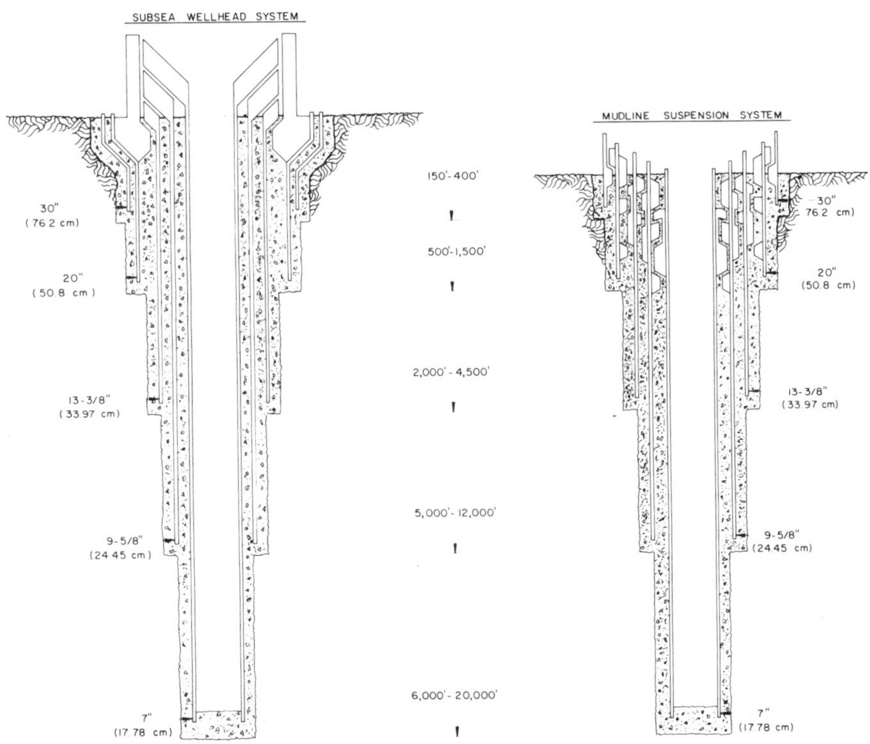

FIG. 3–5 Standard subsea casing program. *(Courtesy Vetco)*

wellhead systems. One such system involving a four-hanger housing is a Vetco 18¾-in., 10,000-psi WP SG-5 single stack to accommodate an 18¾-in. × 13⅜-in. × 9⅝-in. × 7-in. casing program (Fig. 3–6). The 18¾- in. wellhead housing will support the 20-in. surface casing and provide landing and sealing areas for the successive casing hangers and packoff assemblies. The housing also provides the structural support on which the BOP stack (and, subsequently, the completion tree) will be landed, connected, and sealed. The function of the casing hangers is to hang off and support the load of the casing strings within the subsea wellhead housing.

With this housing, a guide base system, consisting of a temporary guide base (TGB) and a permanent guide structure (PGS), will provide an anchor for the guidelines and serve as the initial support structure for installing the wellhead system.

The first piece of subsea equipment used in a typical floating drilling operation is the TGB, through which the drilling assembly will be guided to drill the 36-in. diameter conductor hole. The temporary guide base has four

FIG. 3–6 Temporary guide base. *(Courtesy Vetco)*

guideline attachments on a 6-ft radius (Fig. 3–7). Its four anchor spikes will penetrate 18 in. of the sea floor. The center of this TGB is a funnel-shaped landing ring that accepts the gimballed base of the permanent guide structure, even in soft-bottom conditions.

FIG. 3–7 Permanent guide base and wellhead housing on TGB. *(Courtesy Vetco)*

The permanent guide structure, adaptable for drilling or jetting, supplies the landing seat for the 30-in. wellhead housing. It also provides final guidance for drilling tools, BOP stack, and completion tree if necessary. The gimballed interface between the PGS and TGB permits landing when misaligned up to 10° horizontally and ensures correct alignment of the guide posts and conductor housing during installation. The 10-ft guide posts on 6-ft radius may be fitted with a special post top for remote guideline connection.

Guidelines are trapped at the top and bottom of each post. Tension to maintain parallel guidelines is less critical in shallow-water installations. But for installation in, say, 1,000 ft of water, it is recommended that each guideline be adjusted and hydraulically maintained at 2,000 lb plus the weight of the lines for running tool operations and BOP or (tree) installation. This design provides for replacement of broken or damaged guidelines by diver or remote operation.

In actual operation, the TGB is set on the spider beams of the drilling vessel for attaching the guidelines (usually ¾- or ⅞-in. diameter wire rope). Weight material (barite or cement) is added to the TGB to increase its weight to between 25,000 and 30,000 lb. A running tool is installed in the four J-slot preparations in the TGB with left-hand rotation. The temporary guide base is then run to the ocean floor on drill pipe. The running tool is released by slacking off the weight of the drill pipe and turning the tool to the right ⅛ of a turn. (In the past, each guideline was tensioned with a fixed weight over a sheave. Today, hydraulic or pneumatic guideline tensioners, each with 16,000-lb capacity, keep the guidelines in tension.)

The next operation is to drill the 36-in. hole for the 30-in. conductor casing. The 36-in. drilling assembly is made up and lowered through the rotary table and down through the moonpool or spider deck area.

The next piece of equipment to be prepared is the four-post permanent guide structure. The PGS is set on the spider beams and the four guidelines are inserted and trapped in the guide post slots. The 30-in. casing is then run through the center hole of the PGS.

The first joint of 30-in. conductor (called the *shoe joint*) has a nonreturn or float valve included in the guide shoe that is usually welded to the bottom of the shoe joint. Soft line rope is used often to guide the 30-in. casing shoe into the predrilled hole. The individual 30-in. conductor joints are usually 40 to 50 ft long. The joints of conductor pipe are connected by threadless mechanical connectors.

Each 30-in. casing joint may have pad eyes attached to support the conductor joint in the rotary table during running and connection. After the next conductor joint is stabbed and made up to the joint in the rotary table, the pad eyes are cut off and the casing is lowered to the next set of pad eyes. Thirty-in. casing elevators are sometimes used in lieu of the pad eyes as a means of handling the 30-in. conductor.

After the casing string is made up to the 30-in. housing joint, the conductor string assembly is lowered through the rotary table and landed in the PGS, which has already been set on the spider or moonpool beams below the rotary floor and through which the 30-in. casing joint has been lowered. The 30-in. housing joint is handled using a housing running tool. Lockdown is accomplished by bolting down a split lock plate that secures the 30-in. housing to the PGS. The four vertical posts on the permanent guide structure guide subsequent tools and stabilize equipment for the BOP stack when it is run and landed on the subsea wellhead.

Once a casing string is run and landed in place, circulation is established to clean the hole. Then cement slurry is pumped, under pressure, down through the casing, through the casing shoe, and up into the casing/hole annulus. The liquid cement inside the casing string is displaced with a calculated volume of water or drilling fluid to place the cement in the correct position where it sets (becomes solid) in approximately three to four hours. Usually the 30-in. and 20-in. casing/hole annulus is cemented back up to the mud line. Besides sealing off the upper formation zones, the cement provides a strong pile section to support the heavy weight of the BOP stack when it is attached to the wellhead.

If formation conditions are favorable, the 13⅜ in., 9⅝ in., and 7-in. casing annuli may be cemented up only a few hundred feet inside the shoe of the previous casing string. This, of course, increases the amount of recoverable casing when and if the well is abandoned. The composition and formation pressures of the various zones encountered, as well as government regulations, determine exactly how much cement is set in each annulus.

The set cement seals each annulus to prevent migration of gases or fluids to other zones of the subsoil geologic structure. Cement is also used to shut off highly permeable zones (potential lost-circulation zones), high-pressure zones, or other problem zones. Because of these possible problems, the weight of the cement slurry must be controlled closely to avoid placing large hydrostatic loads on the formation. Cementing protects the casing against corrosion from subsurface mineral waters and electrolytic action from the outside. Cementing also helps strengthen casing resistance to burst pressures.

Tables 3–2 and 3–3 outline some of the primary properties for various casing grades and sizes that are used offshore. Note that J-55 pipe has identical data to K-55, the only difference being that K-55 is rated for hydrogen sulfide gas service and J-55 is not.

In the steel mill, each piece of casing is *drifted,* an internal measurement that indicates the guaranteed minimum inside diameter of the joint. Drift diameter is less than the calculated inside diameter and is an important consideration when using downhole tools. Line pipe usually is not drifted unless this operation is required and negotiated with the pipe manufacturer.

The maximum internal pressure for a particular casing string is given as the minimum internal yield pressure in pounds per square inch (psi). If this pressure is exceeded, the pipe will not retain its original form and rated strength. The well pressure should never exceed the minimum internal yield of the innermost casing string. Collapse resistance and internal yield pressure are critical factors when high pressures migrate into the annulus between two casing strings. If this occurs, either the inner casing string may collapse or the outer string may burst if the corresponding pressure ratings are exceeded. The two most common occurrences causing this type of problem are when high-pressure fluids or gases channel a poor cement job or when high pressure bypasses the casing hanger packoff seal. When casing is used to support a load in tension, the indicated minimum body yield strength represents the maximum allowable load.

Line pipe normally is used for 30-in. and 20-in. casing strings and marine riser rather than high-strength casing. Usually each casing string will be supported at the mud line. One exception is the practice of hanging the 7-in. casing on a liner hanger inside the 9⅝-in. casing. This reduces the amount of 7-in. casing required.

As an alternative to drilling the 36-in. diameter conductor casing, casing may be jetted into place. This procedure establishes the 30-in. casing, permanent guide structure, and the guidelines in one operation and does not necessarily require the use of the temporary guide base. Soft-bottom conditions are prerequisites to using the jetting method.

The 30-in. housing and 30-in. open-ended conductor is made up, run, landed, and locked in the PGS, which sits on support beams as discussed

Table 3–2 API Minimum Performance Properties of Casing

Size (OD) in.	Nominal weights threads & coupling lb/ft	Steel grade	Wall thick-ness in.	Drift diam-eter in.	Col-lapse resis-tance PSI	Pipe body yield strength plain end 1000 lbs.	Internal yield pressure PSI	
							Round thread	Buttress thread
7″	20.00	H–40	0.272	6.331	1,980	230	2,720	
	23.00	K–55	0.317	6.241	3,270	366	4,360	4,360
	26.00	K–55	0.362	6.151	4,320	415	4,980	4,980
	23.00	C–75	0.317	6.241	3,770	499	5,940	5,940
	35.00	C–75	0.540	5.795	9,710	763	8,660	7,930
	29.00	N–80	0.317	6.241	3,830	532	6,340	6,340
	35.00	N–80	0.540	5.795	10,180	814	9,240	8,460
	23.00	C–95	0.317	6.241	4,150	632	7,530	7,530
	35.00	C–95	0.540	5.795	11,640	966	10,970	10,050
	35.00	P–110	0.540	7.656	13,010	1,119	12,700	11,640
	35.00	V–150	0.540	7.656	16,230	1,526	17,320	15,870
$9^5/_8$″	36.00	H-40	0.352	8.765	1,740	410	2,560	
	36.00	K–55	0.352	8.765	2,020	564	3,520	3,520
	40.00	K–55	0.395	8.679	2,570	630	3,950	3,950
	40.00	C–75	0.395	8.679	2,980	859	5,390	5,390
	53.50	C–75	0.545	8.379	11,330	1,166	7,430	7,430
	47.00	N–80	0.395	8.679	5,490	979	5,750	5,750
	53.50	N–80	0.545	8.379	11,760	1,244	7,930	7,930
	40.00	C–95	0.395	8.679	5,910	1,088	6,820	6,820
	53.50	C–95	0.545	8.379	13,030	1,477	9,410	9,410
	53.50	P–110	0.545	8.379	14,100	1,710	10,900	10,900
	71.80	V–150	0.750	7.969	19,640	3,137	18,060	18,060
$13^3/_8$″	48.00	H–40	0.330	12.556	770	541	1,730	1,730
	54.50	K–55	0.380	12.459	1,130	853	2,730	2,730
	61.00	K–55	0.430	12.359	1,540	962	3,090	3,090
	60.00	K–55	0.480	12.259	1,950	1,069	3,450	3,450
	72.00	C–75	0.514	12.191	2,590	2,590	5,040	5,040
	98.00	C–75	0.719	11.781	5,720	2,144	6,270	6,120
	72.00	N–80	0.514	12.191	2,670	2,670	5,380	5,380
	98.00	N–80	0.719	11.781	5,910	2,287	6,680	6,530
	72.00	C–95	0.514	12.191	2,820	2,820	6,390	6,390
	72.00	P–110	0.514	12.191	2,880	2,596	7,400	7,400
	72.00	V–150	0.514	12.191	2,880	3,323	10,090	10,090
16″	65.00	H–40	0.375	15.062	670	736	1,640	1,640
	75.00	K–55	0.438	14.936	1,020	1,178	2,630	2,630
	84.00	K–55	0.495	14.822	1,410	1,326	2,980	2,980
	109.00	C–75	0.656	14.688	2,980	2,372	5,380	
	109.00	N–80	0.656	14.688	3,080	2,530	5,740	
$18^5/_8$″	87.50	H–40	0.435	17.567	630	994	1,630	1,630
	87.50	K–55	0.435	17.567	630	1,367	2,250	2,250
	87.50	X–52	0.435	17.567	630	1,290	2,110	2,110
20″	94.00	H–40	0.438	18.936	520	1,077	1,530	1,530
	94.00	K–55	0.438	18.936	520	1,480	2,110	2,110
	106.50	K–55	0.500	18.812	770	1,685	2,410	2,410
	133.00	K–55	0.625	18.542	1,500	2,125	3,060	3,060

Table 3–3 Line Pipe Dimensions, Weights, and Standard Test Pressures

Size (OD), in.	Weight, lb/ft	Wall thickness, in.	Inside diameter, in.	Standard test pressure, PSI		
				X-42	X-52	5L Std. Grade B
20″	91.51	0.438	19.124	1,660	2,050	920
20″	104.13	0.500	19.000	1,890	2,340	1,050
20″	129.33	0.625	18.750	2,360	2,750	1,310
22″	114.81	0.500	21.000	1,720	2,130	950
22″	142.68	0.625	20.750	2,150	2,500	1,190
22″	170.21	0.750	20.500	2,500	2,500	1,430
24″	125.49	0.500	23.000	1,580	1,950	880
24″	156.03	0.625	22.750	1,970	2,300	1,090
24″	186.23	0.750	22.500	2,300	2,300	1,310
30″	309.72	1.00	28.000	2,520	3,000	1,400

previously. A special jetting tool or bit is made up on a drill pipe stinger that is spaced out inside the 30-in. diameter conductor to position the jetting tool about 10 to 18 in. from the bottom of the 30-in. diameter open-ended shoe joint. A stabilizer is positioned above the jet sub to centralize it inside the 30-in. casing. The plugs in the running tool are removed, allowing the jetted returns to rise inside the casing and spill out of these ports onto the ocean floor. Sea water is used as the jetting fluid.

The 30-in. casing string with permanent guide structure and guidelines are run until the shoe joint reaches the mud line. As the formation is washed by the action of the jetting nozzle, the casing is lowered slowly into the resulting cavity. This procedure is continued until the PGS is a few feet above the mud line. With the assembly in its final position, the drillstring is rotated to the right to release the housing running tool. The jetting assembly, with the housing running tool, is then retrieved. Formation friction on the 30-in. casing is sufficient to anchor and hold the casing in place. The 30-in. casing is considered installed at this point. Recently, 30-in. casing strings have been drilled or jetted into position using dynadrills or turbodrills (downhole drill motors) within the casing string instead of the more conventional jetting tool described above.

The 30-in. casing, usually set to a depth of 80 to 300 ft, provides only structural support and will not withstand pressure. The depth of the 30-in.

casing is determined by the ability of the soil to support the wellhead and other equipment; vertical loading and overturning moment are the criteria. If riser is used when drilling hole for the 20-in. surface casing, the ability of the formation to withstand the hydrostatic pressure of the mud column in the riser must also be taken into account.

For the 20-in. (or surface) casing, a pilot hole is drilled and then opened to 26 in. This casing normally is set to about 1,000 ft below the mud line. In the past, holes for the surface casing were drilled with mud and cutting returns to the sea bed. However, shallow gas sands have caused blowouts, so risers with diverter systems are now used. Diverters are low-pressure annular blowout preventers used to direct the flow of fluids away from the rig floor. After drilling, the riser is pulled because it is too small to accept the 20-in. casing connectors. Then the wellhead and casing are run and cemented with returns to the sea floor. While the cement sets, the riser is again run with the BOP stack. With the surface casing sealing off the relatively low-pressure formations, the higher pressure formations (usually 3,000 ft or more below the mud line) can be controlled.

Use of an 18¾-in. wellhead and BOP stack permits the 17½-in. hole for the 13⅜-in. casing to be drilled without underreaming, eliminating considerable drilling time. After the casing has been cemented and sealed, holes for the remaining casing strings are drilled. Each casing is run, cemented, and sealed. The depth for each casing string is based on the fracture gradient at the shoe of the surface casing, estimated formation pressure at setting depth for the next casing, and the estimated pressure gradient of the fluid in the open hole. Pressure and structural integrity are estimated using the same philosophy used on land.[1]

Blowout Preventer Stack

The blowout preventers maintain control over potential high-pressure conditions that may exist in the formation being drilled. The BOP can shut in the well under pressure while formation material, fluids, or gases that have invaded the well bore are circulated out. Standard sizes and pressure ratings for subsea BOPs are 16¾-in., 5,000 or 10,000 psi; 18¾-in., 10,000 or 15,000 psi; and 21¼-in., 10,000 psi for a single stack system. Subsea wellhead systems, the BOPs, and the wellhead itself generally have the same nominal ID and pressure rating. The BOP components are integrated into a structural steel framework with four stack posts. This assembly, then, is referred to as a BOP stack, BOP, or the stack.

A typical BOP stack consists of a wellhead connector housed in the lower framework; three pipe rams, a blind shear ram, and the annular preventer are encased in the middle framework. Of the four ram preventers, the blind shear usually is placed just below the annular preventer or is positioned so there is one pipe ram between it and the annular preventer. These components are bolted or clamped together before being integrated into the BOP stack framework. A mandrel is installed atop the bag-type (annular) preventer for attaching the lower riser package (Fig. 3–8).

The lower, middle, and upper support frames are flanged and bolted to four large guide posts through which guidelines are run. When the stack is landed on the ocean floor, these stack posts fit over the posts on the PGS. Intermediate landing pads are sometimes located on the posts at the level of the upper stack frame to support the stack on the moonpool or spider beams while the lower riser package is connected. Hydraulic connectors are included, top and bottom, to connect the stack remotely to the subsea wellhead housing and to attach the lower riser package to the stack prior to running.

The pipe rams are operated hydraulically, when necessary, to close and seal around the drill pipe. Their normal operating pressure is 1,500 psi. Cameron pipe rams are equipped with hydraulic wedge locks to hold the rams in closed position. The wedge locks must be hydraulically retracted to open the pipe rams again. Shaffer and Hydril rams have an automatic lock incorporated in their operating mechanisms.

The hydraulically operated blind shear rams can cut through the drill pipe to seal off the BOP stack bore in an emergency. These rams also operate at 1,500 psi but usually have larger operating mechanisms than the pipe rams. The blind shear rams have a locking feature that must be released to open them again.

The annular or bag-type preventer is basically a rubber cylindrical element with bonded metal ribs. When hydraulically actuated, a large piston compresses the rubber element, which then closes on the drill pipe, casing, or itself. The metal ribs help it return to the fully open position. Standard operating hydraulic line pressure for annular preventers is 1,500 psi. Some types of annular preventers must close against the hydrostatic pressure in the riser column; therefore, a pressure-balance system is incorporated for counteraction. Annular preventers used in offshore operations are normally rated at 2,000, 5,000, or 10,000 psi working pressure.

[1]Riley Sheffield, "Floating Drilling: Equipment and Its Use," *Practical Drilling Technology, Vol. 2*, (Houston: Gulf Publishing Company, 1980).

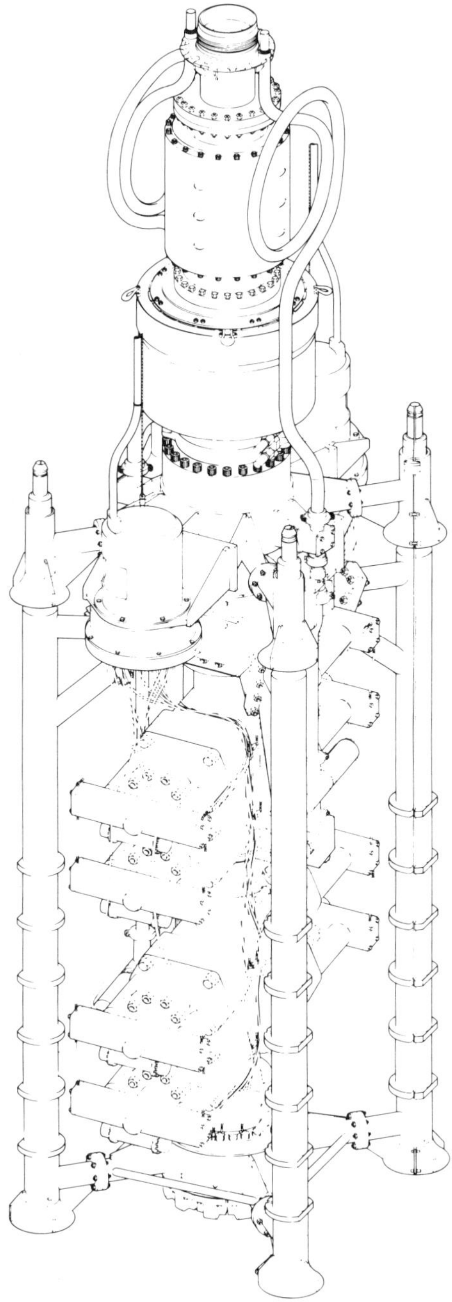

FIG. 3–8 BOP stack with lower marine riser package. *(Courtesy Vetco)*

The hydraulic connectors used between the wellhead and the BOP and between the BOP and the riser are controlled from the surface and are of two basic types: mandrel and collet. Mandrel-type connectors utilize hydraulically actuated cams to drive locking dogs into the grooves machined into the wellhead (Fig. 3–9). The collet connector uses a series of collet fingers to form a funnel configuration to guide the connector over the well hub (Fig. 3–10). Both types of connectors use AX or VX rings for sealing.

Usually, two choke-and-kill lines are run down to the BOP stack, either integrally with the riser or as independent lines. Choke-and-kill valves provide subsea shutoff of the high-pressure lines and are part of the stack, controlled by the BOP control system. Most BOP stacks have connections below all rams to allow different choke-and-kill hookups, as many operators have specific requirements with regard to the placement of choke-and-kill entry lines within the stack.

Each choke-and-kill line usually carries two fail-safe close (FSC) gate valves near each stack connection. These valves are operated hydraulically. In the event of sudden pressure loss, a spring cartridge in the valve operator automatically moves the valve gate to the closed position. Choke-and-kill valves must be rated for the same working pressure as the BOP stack. The upper terminal fitting for the choke-and-kill lines on the BOP stack are male stab subs that mate with the female connections on the lower riser package.

Because of their relative inaccessibility, subsea BOPs demand redundancy to an extent not required for surface stacks. As a result, numerous combinations and configurations have been tried with mixed success (Fig. 3–11). The trend seems to be toward a single-stack system (as opposed to a two-stack system where minimal low-pressure equipment drills a shallow hole and lower formations are drilled with a smaller diameter high-pressure stack). However, the trend may be reversed if extremely high pressures (greater than 10,000 psi) are encountered frequently in floating drilling.

Of necessity, BOPs require highly efficient hydraulic control systems capable of operating any of the functions on the stack with precision and dispatch. For instance, high flow rates are required for the large volumes needed to operate the ram and annular preventers. These volumes may range from 5 gal for small rams to 50 gal for a large annular preventer.

The NL (Koomey) system best illustrates redundant hydraulic operation (Fig. 3–12). The basis of this hydraulic control system is a male control pod connected to the control receptacle mounted on the upper frame of the BOP

FIG. 3–9 Vetco H-4 connector. *(Courtesy Vetco)*

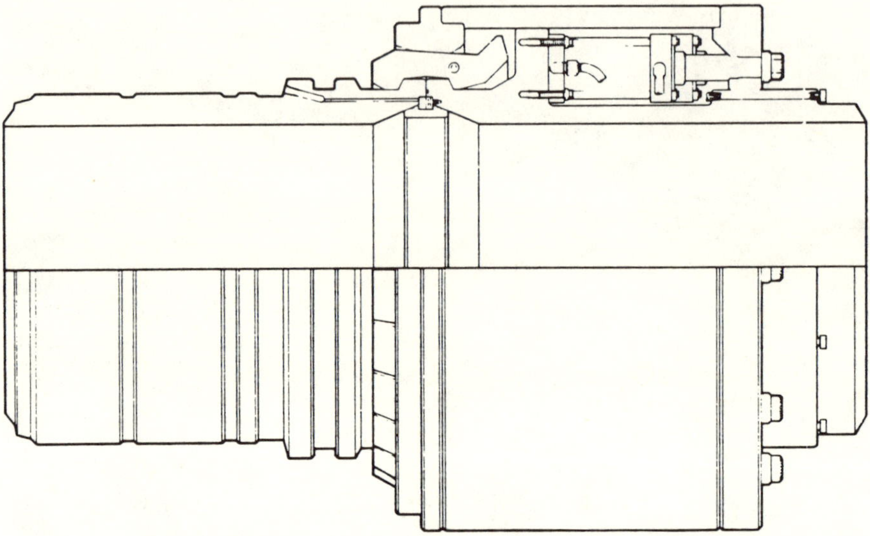

FIG. 3–10 CIW collet connector. *(Courtesy Vetco)*

stack. This control pod houses the SPM valves that direct the hydraulic fluid to operate the various functions on the stack. The mating surface of the control pod has outlets with individual seals that match inlets on the conical sealing face of the female control receptacle. The female receptacle has outlets that usually are connected by flexible high-pressure hose to the various functions of the BOP stack.

To provide 100% redundancy in the BOP control, an identical control system is installed on the stack 180° from the other control pod and receptacle. In the event of hydraulic control failure through one hose bundle and control pod, the other system may be actuated. Hydraulic power can be directed to a function from either control pod through shuttle valves mounted on the operating functions of the BOP. Two inlets to the shuttle valve are connected to the hoses from the female receptacles of the hydraulic control system. Inside the body of the shuttle valve, a sealing shuttle piston moves with operating pressure from inlet to inlet, directing the hydraulic fluid to the outlet connected to the function.

When a BOP stack employs a single female system for control, the female receptacles are mounted on the upper frame of the BOP. The male control pods lock and seal in the female receptacles and are run with the stack. The individual control pods and hose bundles of a single female system can be

FIG. 3–11 BOP stack in yard. *(Courtesy Vetco)*

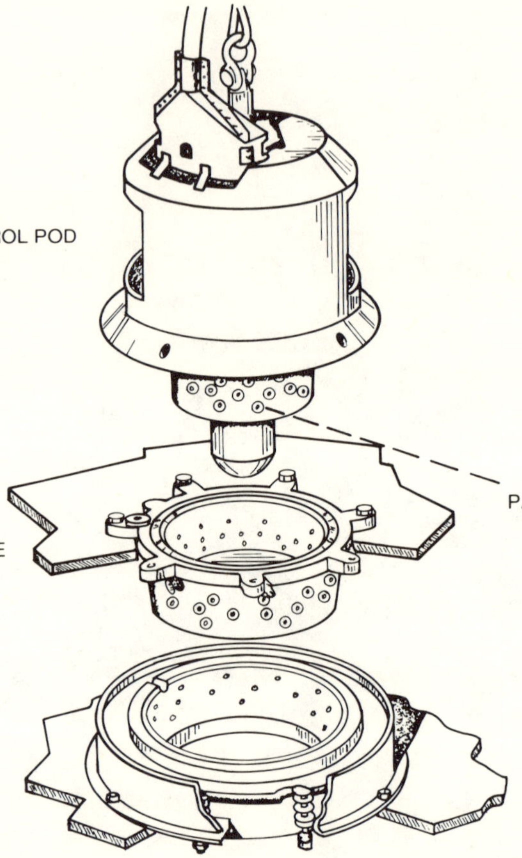

FIG. 3–12 Koomey system for redundant hydraulic control. *(Courtesy Vetco)*

retrieved independently and rerun to the BOP stack with the use of a small guide frame that is run on two guidelines. To release the riser from the stack in an emergency situation, the riser connector and both control pods must be released.

The support equipment for the hydraulic control system consists of two power-driven hose reels that spool off and rewind the control hose bundle, a master control panel with hydraulic pumps and fluid supply reservoirs, and a driller's control panel. Often a mini-panel is included in the toolpusher's office. The regulated output pressure from the master panel is usually 1,500 psi with 3,000 psi available.

It is now standard practice to install a rack of accumulator bottles on the BOP stack to provide a sufficient volume of hydraulic fluid to operate all functions on the stack. This avoids the slow reaction time that would result if all the required fluid had to be transferred through the hose bundle down to the BOP.

The Koomey double female system for BOP control is comprised of a male control pod connected to the hydraulic control hose bundle that houses the hydraulically controlled SPM. As in the single female system, the SPM directs the main hydraulic power to the various functions on the stack. In the double female system, the control pod seals and locks in an intermediate female receptacle bolted to the lower riser package. When the lower riser package is installed on the BOP stack, the male mated to the intermediate female receptacle lands and seals in the female receptacle in the upper frame of the stack.

BOP stacks are always equipped with two control pods for complete backup hydraulic control. Hydraulic pilot hoses and a hydraulic supply hose make a bulky package run from the surface—moreso as water depth increases. For this reason, a number of manufacturers are now providing electrohydraulic control systems.

With an electrohydraulic control system, electrical power, fed through a multicore conductor, actuates solenoid-operated valves in the control pod. Then, only a single hydraulic hose is required to deliver the hydraulic fluid to the accumulators. Some systems have incorporated the accumulator fluid supply line into the integral riser system.

Pure electric control or multiplex systems are also being manufactured where electric signals and electrical power are sent to the BOP stack through a single, multicore conductor connected to each control pod. Acoustic control systems generally are not being considered as a prime operating system but are being employed as a backup control system for subsea BOPs.

If and when a higher-then-expected well bore pressure is encountered while drilling, the annular and ram preventers can be closed to seal around the drill pipe. The pressure buildup is controlled by circulating heavier drilling mud through the drill pipe and back through the choke or kill lines. Once the mud column pressure exceeds the formation pressure, the preventers can be opened and drilling may continue.

In the event sea conditions make it imperative to disconnect from the BOP stack, an emergency drill pipe hangoff tool can be landed in the subsea wellhead housing. The well can then be sealed by closing the bottom pipe ram around the drillstem of the emergency drill pipe hangoff tool. The running tool is released from the hangoff tool, and the remaining drill pipe may be retrieved. At this point, the blind ram is closed above the hangoff tool and the hydraulic connector in the lower riser package is disconnected. The drilling vessel is then in a position to ride out the emergency, leaving the well in a secure and shutin condition. During the reconnection and reentry procedure, it is important to circulate through the choke-and-kill lines so any pressure built up under the closed rams is bled off before the drill pipe is lowered and reconnected.

Only under extreme emergency conditions are the blind shear rams closed on a drillstring. In such case, one of the pipe rams below the shear rams preferably would be closed first, just below a drill pipe tool joint. In this way, the well can be circulated through the choke-and-kill lines before the blind shear rams are reopened. The cut section of the drill pipe is in the sealed area between the blind shear ram and the pipe ram. The sheared end of the drill pipe is open enough to allow mud from the kill line to enter the drill pipe and be circulated to the surface through the choke line below the closed pipe ram.

Since the BOPs are the primary means of controlling the well in the event of a kick, the importance of their proper operation and maintenance cannot be overemphasized. All rig personnel should be familiarized with BOP operation—their lives could depend on it.

Thorough testing and maintenance of the BOPs must be conducted on the surface to ensure a long and dependable operational life subsea. For surface testing, a test stump carrying the male profile of the wellhead provides a landing area for the BOP stack. When the hydraulic connector at the bottom of the stack is latched onto the stump, all seals may be checked for leaks. The following should be observed when surface-testing the BOP:

1. Any special diagnostic procedures recommended by the manufacturer for each component should be conducted.
2. The control system should be pressure tested before testing the BOP.
3. All components should be tested at rated working pressure.
4. Pipe rams and annular preventers should be closed around drill pipe to avoid damaging the seals.
5. All hydraulic connections and connectors must be inspected carefully for leaks.

In pressure testing the BOP stack with a drill pipe test tool installed in the test stump, it is advisable to leave the bore of the test tool open to the atmosphere so leaks may be detected quickly. The open bore also prevents the test tool from becoming a projectile in the event of thread failure because

FIG. 3–13 BOP stack being installed from semi. *(Courtesy Vetco)*

the test pressure acts across the effective area of the drill pipe cross section and considerable upward force is generated.

At the conclusion of the functional and pressure testing, the BOP stack is disconnected from the test stump and is ready to be run on drill pipe running tools or on the marine riser (Fig. 3–13).

Subsea testing, of course, cannot be as extensive as surface testing. The frequency of subsea testing may vary with operators as well as with local regulations governing the minimum frequency of subsea testing. A test plug run on drill pipe seals off the casing while testing. Flow from the pipe would indicate a leak in the test plug. The blind shear rams may be function-tested periodically, but no satisfactory method is available currently for pressure testing without high risk of rupturing the casing.

Drilling Riser

The marine (or drilling) riser is the communications link between the drilling vessel and the subsea wellhead through which downhole equipment is guided and mud is returned to the surface. It also serves as a running string for the BOP stack. Improvements in materials and construction have greatly reduced the probability of riser failure, but proper care and use is still required to reduce riser deterioration.

Without vessel motion, the marine riser would be just a line of pipe like those used in land operations. Because of these motions, however, the marine riser becomes a more complicated system, requiring additional special equipment. In more complex and hostile offshore environments, drilling risers must be designed to withstand the maximum combinations of riser tension, internal pressure, and bending moments without overstress or fatigue. The physical demands imposed by drilling vessel excursions; wind, current, and wave forces; and thermal expansion require extremely reliable components.

A typical riser system is composed of a lower marine riser package, riser joints with high-strength connectors, a telescopic joint with terminal fittings and a tensioning ring, and a diverter. Of course, a tensioning system must be incorporated and, in some cases, buoyancy modules. Some risers also include a fillup valve that prevents riser collapse in the event of drilling-fluid pressure loss. Sometimes, in deep-water drilling, an upper ball (or flex) joint is used with the telescopic joint. However, the necessity of the additional ball joint is now being questioned since wells have been drilled in over 2,000 ft of water without an upper ball joint. The choke-and-kill lines may be attached integrally to the riser.

The lower marine riser package is a preassembled unit that connects the marine riser to the BOP stack. Usually it is composed of a lower marine riser guide frame enclosing a hydraulic connector, an annular preventer, a flex joint, and choke-and-kill flex lines. If a double female hydraulic control system is being used, the guide frame also houses the intermediate female receptacles, which in turn carry the control pods.

The flex joint in the lower marine riser package accommodates up to 10° deflection from the vertical to allow for any horizontal movement to which the drilling vessel might be subjected. When using a ball joint, the weight of the riser and the drilling fluids within it creates an unbalanced force on the socket section, tending to force it down onto the ball section. This compression force is balanced by pressurized lubricating oil between the top socket section and the ball section. The hydrostatic head of drilling fluid in the marine riser and the overpull of the riser tensioners creates an upward (or tension) force on the socket section, acting on the lower face of the ball section. These upward forces are variable and are again compensated by pressurizing the lubricating oil between the ball section and the lower socket section. A single ball flex joint pressure-balance system is shown in Fig. 3–14. Hydraulic fluid from the BOP stack control system is applied to the base of the floating piston in another fluid-oil separator. This transfers the pressure into the lubricating oil, maintaining the required balancing force to compensate for the overpull and mud weight. The required hydraulic pressure is dependent upon the tension load on the riser, water depth, and mud weight.

To avoid twisting the flexible choke-and-kill lines around the flex joint, an antirotation pin is incorporated to prevent the socket from rotating relative to the ball section. Extensive wear on the inner bore of the ball joint can occur as the result of drilling while the vessel is displaced from the well bore or is off location. Inserting a replaceable wear bushing will help avoid irreparable damage to the ball joint. A tensile load capacity of 1,000,000 lb has proven ample for running the larger BOP stacks and for the tensioning loads required of the riser.

Because of the pressure requirements anticipated for a ball joint in deep water (3,000 to 6,000 ft), a nonpressurized flex joint was developed with the high tensile capability to handle the deep-water subsea equipment. Vetco's Uniflex (Fig. 3–15) is an example. Since the Uniflex joint requires no hydraulic balance pressure, its operation is simplified and service and maintenance requirements are substantially reduced. The inner surfaces, subject to drill pipe wear, carry removable bushings. Primary flexing takes place at each of the two bearing rings in the upper and lower sections (Fig. 3–16). The two pieces in the middle, the seal assembly, are composed of the same flexing

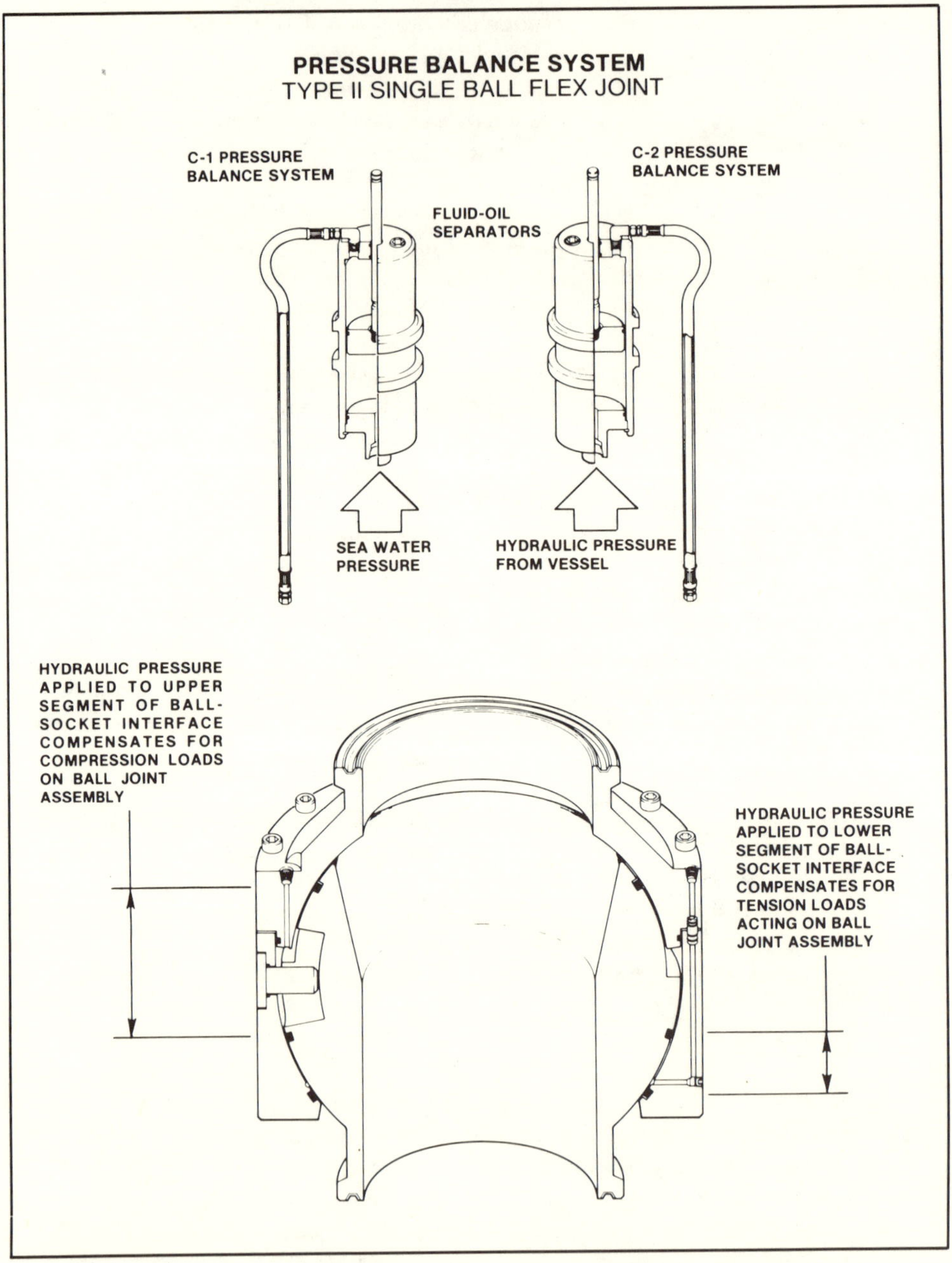

FIG. 3–14 Pressure balance system—Type II single-ball flex joint. *(Courtesy Vetco)*

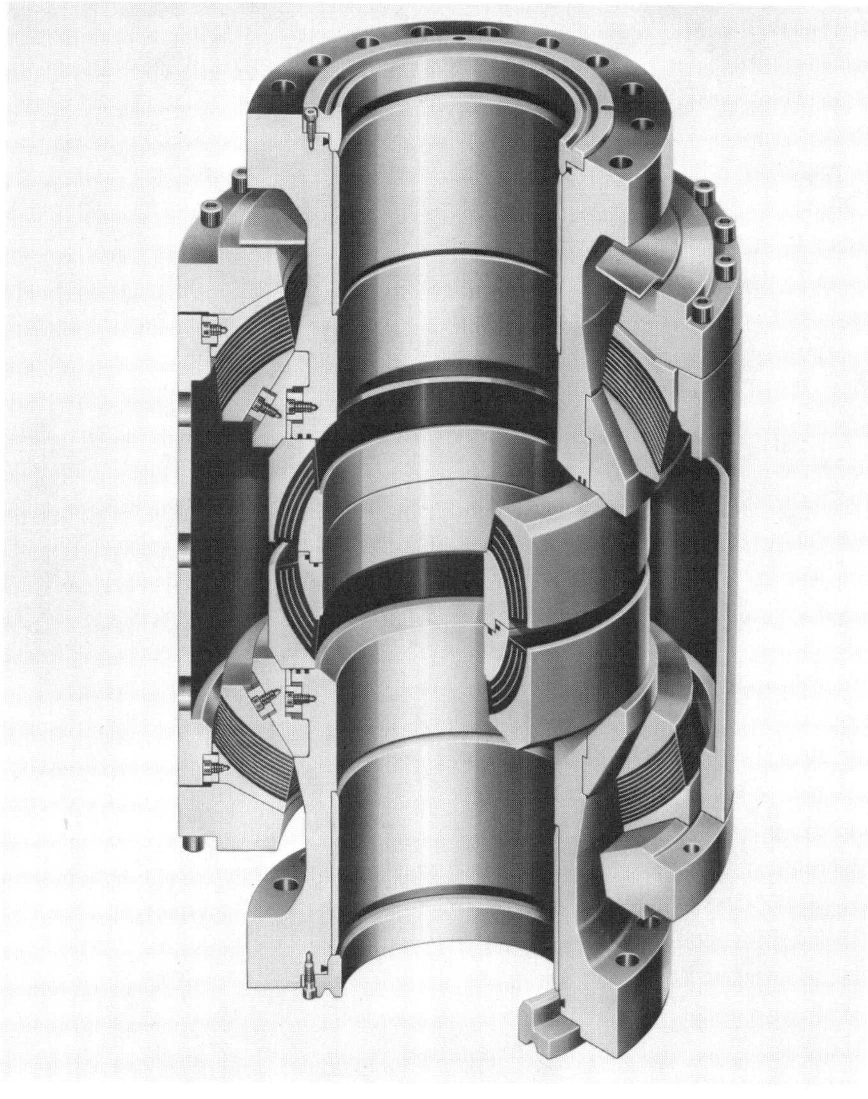

FIG. 3–15 Vetco uniflex joint cross section. *(Courtesy Vetco)*

material and mainly seal between the internal mud pressure and the external ambient pressure. The flex material is laminated layers of steel and rubber. The action is more like a sliding, compressive loading than the pivotal loading on the spherical surface of the single ball joint.

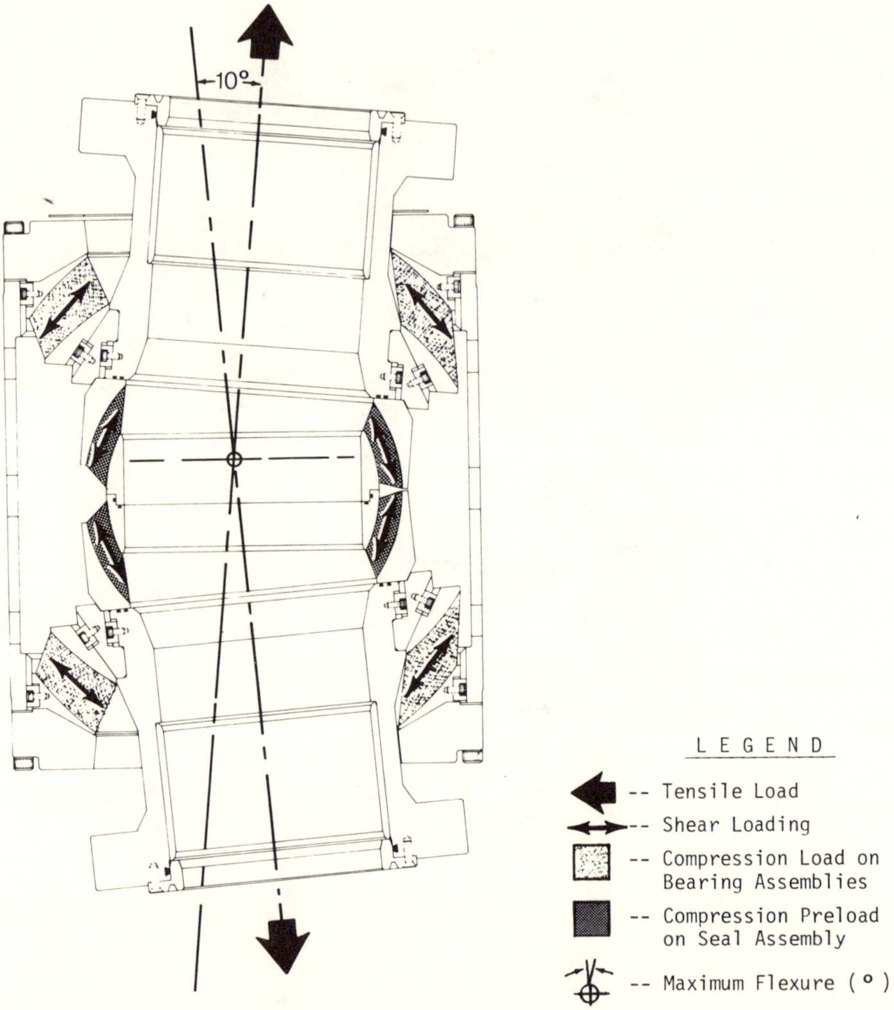

FIG. 3–16 Flexing uniflex. *(Courtesy Vetco)*

The drilling riser joints may be connected by any number of means, including welding, threading, snap rings, segmented locking dogs, and bolted flanges. The requirements for fast makeup and disconnect, rigid pressure-tight connections, minimal stress variance, compactness and light weight for minimum tensioning and buoyancy needs, and easy maintenance must all be considered. Usually, the female (box) portion contains dogs or ring segments, actuated by bolts, that clamp into grooves machined into the male (pin) member of the connector. The connectors are welded on the ends of each riser joint so the riser may be run similarly to drill pipe by stabbing one joint at a time and tightening the connector.

A marine riser connector provides the crossover from the flex joint to the marine riser pipe. This connector is run in a pin-up/box-down mode. It is exactly the same as the connectors used for joining all subsequent riser joints and the telescopic joint.

An integral marine riser system is the modern approach to running the BOP stack and is the method most commonly used today. With this system, the individual choke-and-kill lines and, in some cases, hydraulic fluid supply lines are attached integrally to each riser joint. Running the BOP stack, riser, and choke-and-kill lines all in one operation represents a significant economical advantage because it takes less time. Also, the choke-and-kill lines can be pressure tested against the closed choke-and-kill line valves on the stack as they are run (Fig. 3–17).

The physical integrity of the riser joints is preserved by eliminating all welding except circumferential welds of the connectors to the pipe ends. The main support flanges for the choke-and-kill lines are fabricated as part of the connector, with clamps for additional support for the lines positioned as necessary on the pipe body.

Ribs to guide the riser through the rotary table are fabricated as part of the support flange, making welding unnecessary—again protecting against areas of stress concentration. To accommodate high tensile and bending loads further, a gradual thickening of the pipe wall is incorporated where it connects to the marine riser connector. Usually, marine riser joints are made from seamless X-52 line pipe material and come in 50-ft lengths. Deep-water marine riser can be furnished in 65-ft lengths. A set of pup joints to space out the riser for any water depth usually consists of four joints: 5 ft, 10 ft, 20 ft, and 25 ft.

The telescopic joint consists of an inner barrel that slides into an outer barrel, compensating for the heave of the drilling vessel. The inner barrel, connected to the vessel by a ball joint or gimbal, allows the vessel to pitch and roll without twisting the riser. The mud flow line and the diverter system are located between the inner barrel and the rig floor. The diverter is a low-pressure annular preventer that seals off the well bore. The diverter redirects the flow during a kick, keeping mud and cuttings from blowing onto the rig

FIG. 3–17 Integral marine riser joint. *(Courtesy Vetco)*

floor. The outer barrel of the telescopic joint supports the riser. Riser tension is maintained by the tensioning system, attached by wire rope to the tensioning ring at the top of the outer barrel.

Marine-riser tensioning systems provide constant tension to support the riser while compensating for wave-induced motion of the floating drilling rig. These systems are best suited to situations where excessive string weight or vessel heave are significant factors, such as with deep-water drilling and under adverse weather conditions. The ideal riser tension has been defined as "the tension that will minimize the probability of damaging the riser or drilling equipment, yet cause minimal wear to the tensioners" in any given situation. In some cases, it is desirable to use buoyancy modules along the length of the riser. Frequently, the buoyancy provided by modules reduces tensioning requirements by a like amount. In addition, the resultant lower drag coefficient afforded by the smooth surface of the modules reduces riser stress. Anything as costly and essential to operation as the drilling riser deserves critical appraisal and evaluation, particularly the tensioning requirements in different situations.

Motion Compensating Systems

The motion compensator isolates the drillstring from wave-induced motions of the floating drilling vessel so drilling operations may proceed as if being carried out from a fixed platform. This means all operations are performed more easily—landing equipment, hole opening, drilling, coring, testing, and logging—and the whole drillstring is subjected to less wear and shock. A reliable motion compensator provides more safety to the drilling operation, making every hour of drilling more productive. A typical motion compensator system consists of four basic assemblies: the motion compensator unit, located between the hook and traveling block; a hydraulic-pneumatic operating system; an air supply system; and a driller's control panel (Fig. 3–18).

The motion compensator unit is comprised of an upper yoke assembly to which the traveling block is attached, a lower yoke assembly to which the hook is attached, single or dual hydraulic cylinder assemblies connecting the upper and lower yokes (Fig. 3–19), a hydraulically actuated locking pin to lock the unit mechanically in its retracted position, a hydraulic lock/deceleration valve assembly and accumulator, and an extensometer to indicate point of cylinder stroke. In addition, a rotary hose assembly is supplied to act as a conduit for all electrical and hydraulic/pneumatic power to and from the unit.

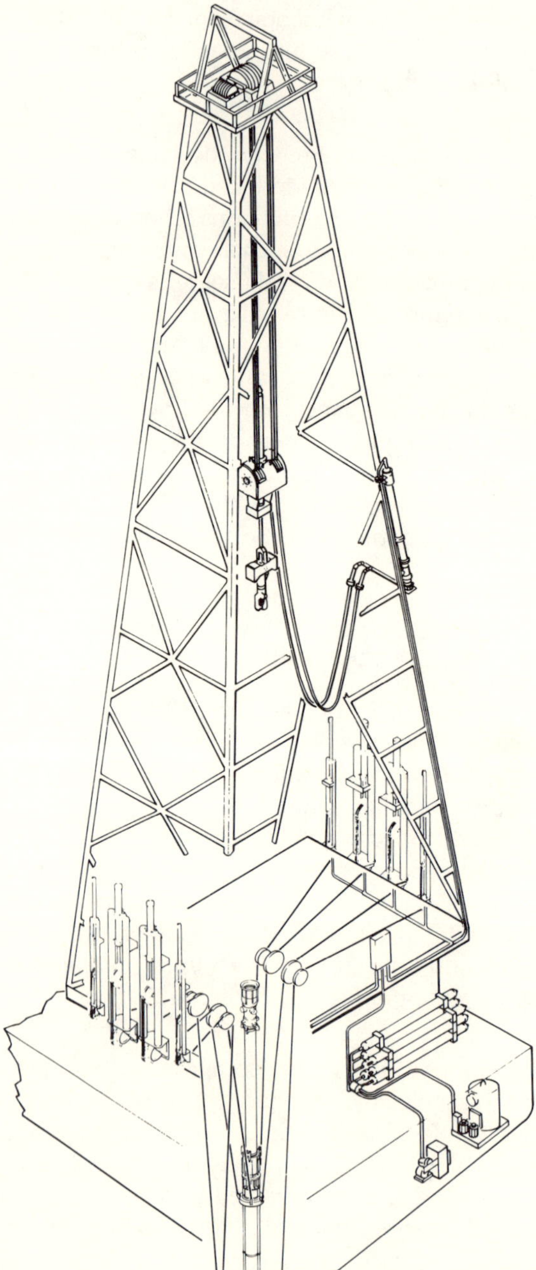

FIG. 3–18 Drillstring motion compensator. *(Courtesy Vetco)*

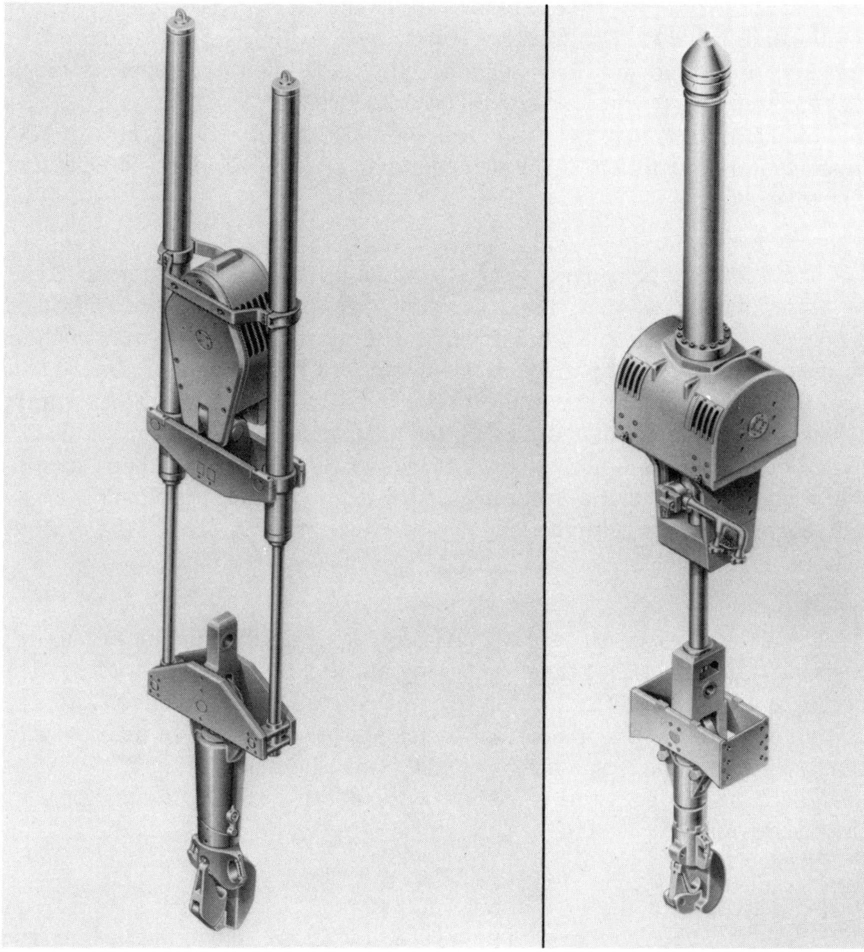

FIG. 3–19 Hydraulic cylinder assemblies—dual and single. *(Courtesy Vetco)*

The single cylinder unit utilizes a large central hydraulic cylinder mounted between two halves of a split traveling block. Variable sheave reaving permits stable and dynamic balance of the unit when operating with 8, 10, or 12 lines. In the design shown, the split block and cylinder are an integral part of the unit, so no other traveling block is required. The hydraulic cylinder is removed easily for repair or replacement on a noninterfering basis, allowing normal drilling operations to continue while the cylinder is absent.

The dual cylinder unit uses an upper and lower yoke assembly that is attached between the traveling block and hook. The yoke assemblies are interconnected by two cylinders that straddle the traveling block. Generally speaking, the dual type is easier to install on an existing rig because the crown and traveling blocks can be utilized and few, if any, derrick modifications are required. However, both types have identical operating characteristics.

The hydraulic/pneumatic system is made up of an accumlator assembly; an accumulator backup system, consisting of a series of air-pressure bottles; a fluid supply unit; and hydraulic fluid. The accumulator assembly cylinder contains a floating piston that acts as an interface between the hydraulic fluid operating the compensator unit and the pressurized air from the air supply system. When vessel heave causes the compensator unit to stroke, fluid is forced in and out of the compensator cylinders and the accumulator, moving the floating piston. If the accumulator runs out of fluid, the floating piston will seat on the bottom, preventing air flow onto the hydraulic hoses—a safety feature.

Usually, the air-bottle assembly will consist of six air-pressure bottles and associated valves, each bottle containing approximately 25 cu ft of air. The bottles are interconnected. Using an air-operated isolation valve, remotely controlled from the driller's control panel, allows up to 60% of the operating bottle volume to be isolated temporarily. This eliminates the need to reduce the entire system pressure to perform operations requiring significantly reduced pressure.

At full capacity, the fluid supply unit storage reservoir contains 395 gal of hydraulic fluid. Also included are two air/oil pumps, a filtering system, and a fluid sensing system. The fluid sensing system automatically maintains the hydraulic fluid level of the accumulator at a safe level by pumping the hydraulic fluid back into the system when required.

It is important to use a hydraulic fluid with excellent physical properties with respect to viscosity and sensitivity to changes in temperatures.

The air supply system should consist of two or more air compressors and associated dryers to provide dry, high-pressure air to the system (3,500 psi). One or more air storage bottles ensure a ready supply of high-pressure air to both the motion compensator and tensioner systems.

Operating controls and indicators are installed on the driller's control panel located on the drilling floor (Fig. 3–20). This enables the driller to operate and monitor the entire motion compensator system. The manually operated controls open and close the valves, adjust the operating system pressure to the operating load, and raise or lower the isolated system pressure. Large, easy-to-read gauges display the various system pressure readings, and indicators show the lock pin position and piston rod(s) extension.

A motion compensator system functions as a hydraulically loaded tension spring with an adjustable tension effort. The compensator's hydraulic operating system is passive. Rather than being driven by hydraulic pumps, the system is energized by a compressed air supply system that acts on a given amount of hydraulic fluid by means of an accumulator. Ideally, the volume of fluid never changes but is displaced back and forth between the compensator and accumulator by the compensator piston. The piston, then, carries the hook load by virtue of hydraulic pressure, maintained by a counterbal-

FIG. 3–20 Motion compensator control panel. *(Courtesy Vetco)*

ancing force induced by air pressure in the accumulator. The pressure is controlled from the driller's panel, allowing rapid variation in effective bit weight.

A combination hydraulic lock/deceleration valve blocks the flow of oil from the compensator cylinder, preventing extension of the piston rods while making connections. This is an important time-saving feature because it eliminates the need to cycle the piston rods fully and double-set the slips. The hydraulic lock/deceleration valve also allows immediate pickup of the drill pipe with the cylinders in any extended position so the drill pipe is held off bottom and motionless with respect to the vessel. In this way, the BOP ram rubbers or bag preventer may be closed quickly and safely around the drill pipe.

In operation, the weight of the drillstring is supported by the hydraulic cylinder assembly positioned between the hook and traveling block. As the drilling vessel heaves with the sea, the piston rods extend or extract to isolate the vertical motion of the vessel from the string. When starting to drill, the operator determines the weight on the hook and simply adjusts the system pressure to balance the hook load. Desired bit load is achieved by reducing the system operating pressure by a sufficient amount for the desired bit weight on bottom (Fig. 3–21).

While running the drillstring, the compensator unit is mechanically locked in its fully closed position. When the last joint of drill pipe has been added to the string, the compensator is unlocked and stroked out (to its fully extended position). The traveling block is then lowered to land the bit on bottom. After tagging bottom, the traveling block is lowered farther to compress the compensator cylinders to the midpoint of their stroke.

A motion compensating system that utilizes a passive operation must rely on a resisting force from the drillstring to indicate that motion has occurred on the ship. This force normally is provided by the weight of the drillstring, the vertical drag of the drill pipe in the hole, and the inertia of the drillstring itself. During wire-line operations (such as logging or testing) there usually is not enough drag and inertial force to cause the compensator to stroke. In these instances, the additional force is provided by a sensing line. The sensing line is run over a sheave hung from the hook and is connected to the drilling riser at one end, attached to the derrick floor at the other end. Tension is then applied to the line by adjusting the compensator system operating pressure. Wire-line operations may then be performed, free from vessel heave (Fig. 3–22).

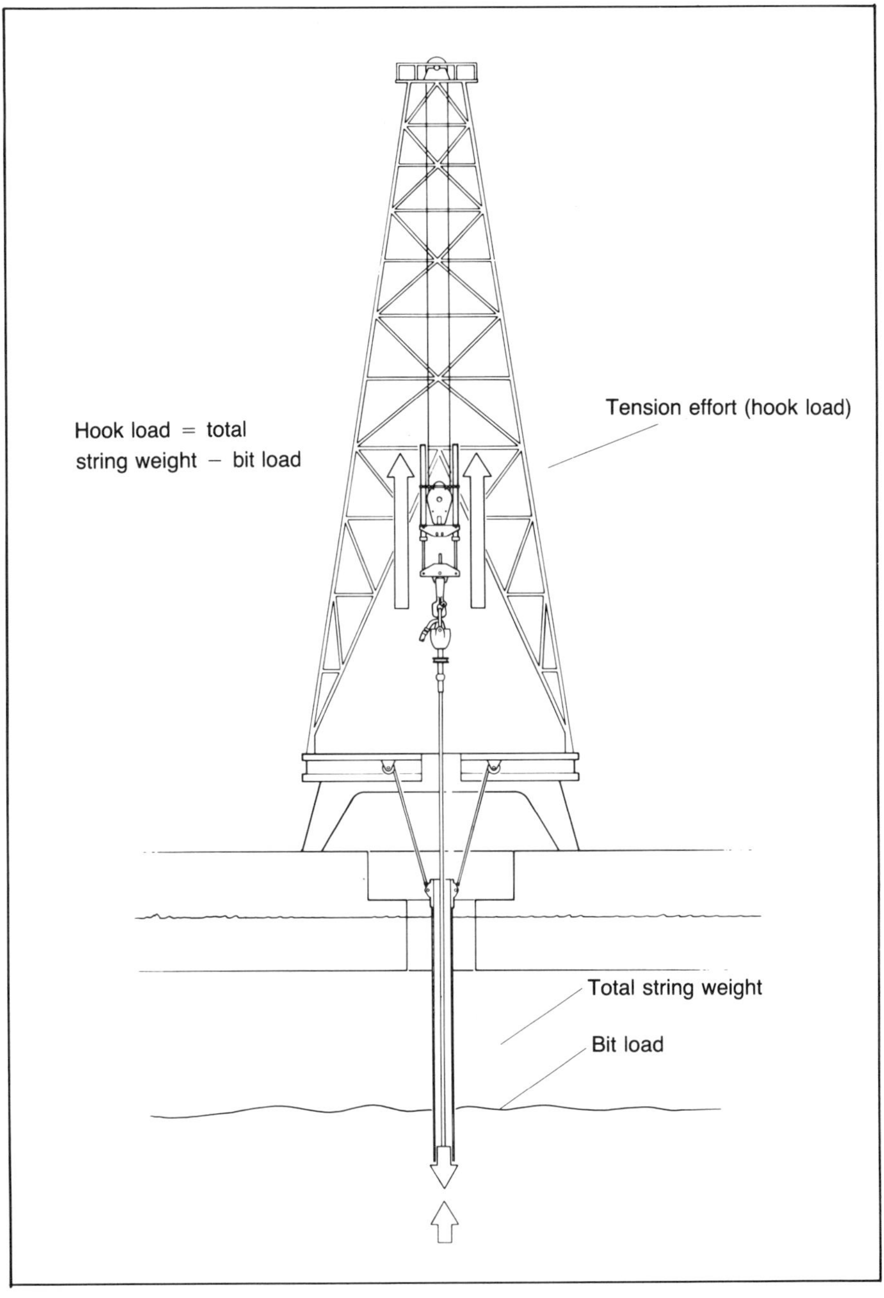

FIG. 3–21 Motion compensator force diagram. *(Courtesy Vetco)*

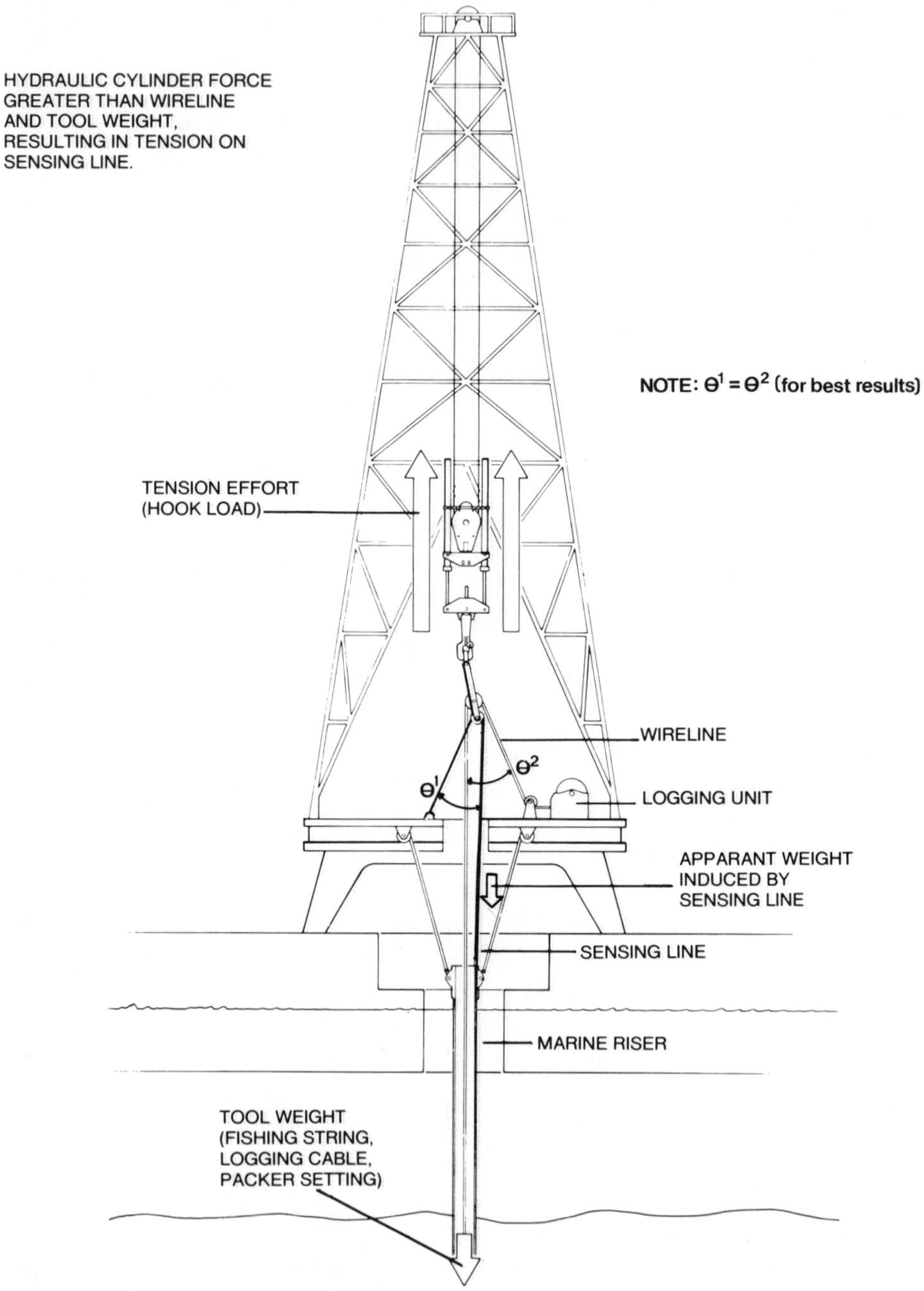

FIG. 3–22 Motion compensator during wire-line operations—simplified diagram. *(Courtesy Vetco)*

4

Tension-Leg Platform Drilling

Although still in the preliminary stages, the tension-leg platform (TLP) concept appears a viable means of affording offshore oil and gas production while remaining relatively economically insensitive to water depth. In many cases, the TLP offers distinct cost advantages over fixed-platform installations, yet it retains many of the platforms' desirable operational characteristics, such as stability, ability to use land-type drilling equipment, and larger overall production capacity than floating facilities.

The offshore industry has been interested in the tension-leg platform for several years. Research firms, manufacturers, users and potential users, and any number of professional organizations and universities have participated in exploring various theories and ideas for TLP drilling, production, and mooring systems. Some having gone as far as designing and testing prototype equipment. System, environment, and platform geometry have been treated specifically to develop a cost-efficient extension of existing technology. Very soon, the world's first TLP will lie securely fixed in 485 ft of water in the North Sea.

By iterative process, a design was selected by Vetco Offshore Inc. as a practical approach to TLP drilling. Detail design and testing of products followed. The result, described here, and depicted in Fig. 4–1 represents current TLP drilling technology and promotes the advancement of the TLP concept in the quest for offshore energy sources.

SYSTEM DESCRIPTION

The unique design of this TLP permits the use of land equipment and modified conventional platform methods, thus allowing simultaneous drilling and production to be safely conducted. The tethers that tie the platform to mooring templates on the ocean floor are kept in tension through the worst weather and demanding deck loading conditions by the buoyancy of the platform's integral columns and pontoons. Deck designs and well arrangement optimize platform space utilization. Use of a high-pressure drilling riser permits blowout preventers (BOPs) to be located in the drilling rig substructure for easier, more reliable installation, operation, and maintenance. Drilling, completion, and production are accomplished without large conductors from subsea wellhead to platform, minimizing weight and reducing stress. The high-pressure drilling riser connectors, modified to accept AX seal gaskets, ensure sealing integrity comparable to hydraulic connectors which attach BOPs to subsea wellheads.

The TLP drilling system consists of a subsea wellhead to which the high-pressure drilling riser is attached; a tapered transition section just above the sea floor to minimize stress and increase fatigue life; drilling riser joints; and a gimballed tensioning system for support. Upon completion, each well will

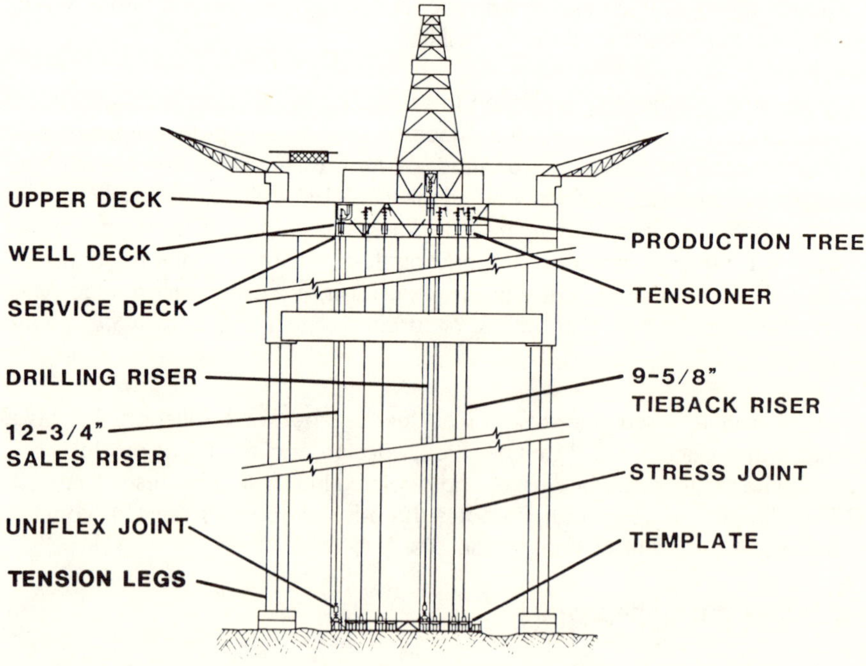

FIG. 4–1 TLP production system. *(Courtesy Vetco)*

be tied back to the platform for control at deck level. Key components are simple and inherently reliable but will have redundant backups and are replaceable for inspection and maintenance.

TEMPLATE

The 21-well unitized template design shown in Fig. 4–2 was based on strength requirements and use of an easily procurable material. Possible installation and leveling techniques are discussed fully in "Floating Drilling."

This template contains 17 standard wellhead receptacles in addition to four leveling pile inserts that may become well slots upon completion of the leveling operations. The 21 wells share guideposts, as they are spaced 8½ ft between centers. Also included at one end of the template structure is a sales riser base/receptacle. The sales line receptacle will receive the 12¾-in. OD pipeline.

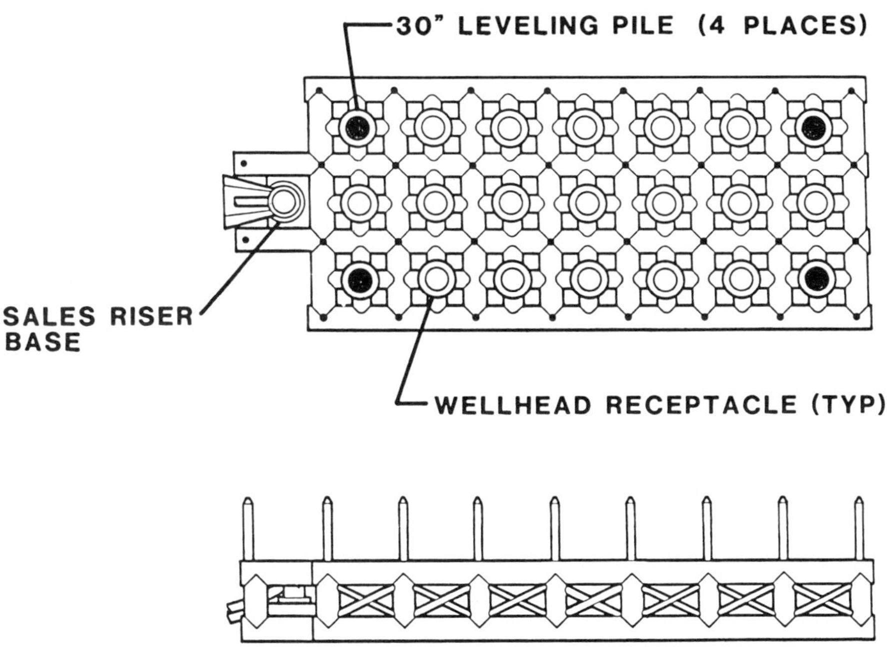

FIG. 4–2 21-well unitized template. *(Courtesy Vetco)*

The wellhead receptacles allow wells to be drilled using standard subsea drilling techniques. During template fabrication, the guideposts for the wellhead receptacles are held in position at top and bottom during welding to ensure verticality and critical spacing. To install the wellhead receptacles, a fixture is run down over the four guideposts of the slot where the receptacle is to be placed. The center of the fixture has an outer profile that matches the inner profile of the receptacle. A tapered section below the receptacle centralizes it on its support beams. When the receptacle is landed and centralized, it is welded to the large support beams of the template.

There is likely to be quite a time lag between setting the template and completing the TLP; consequently, a number of preliminary operations are carried out by a floating drilling vessel (Fig. 4–3) including:

1. Setting and leveling the unitized template
2. Connecting the sales pipeline
3. Setting the anchor template and piles for mooring the TLP
4. Drilling a number of the subsea template wells

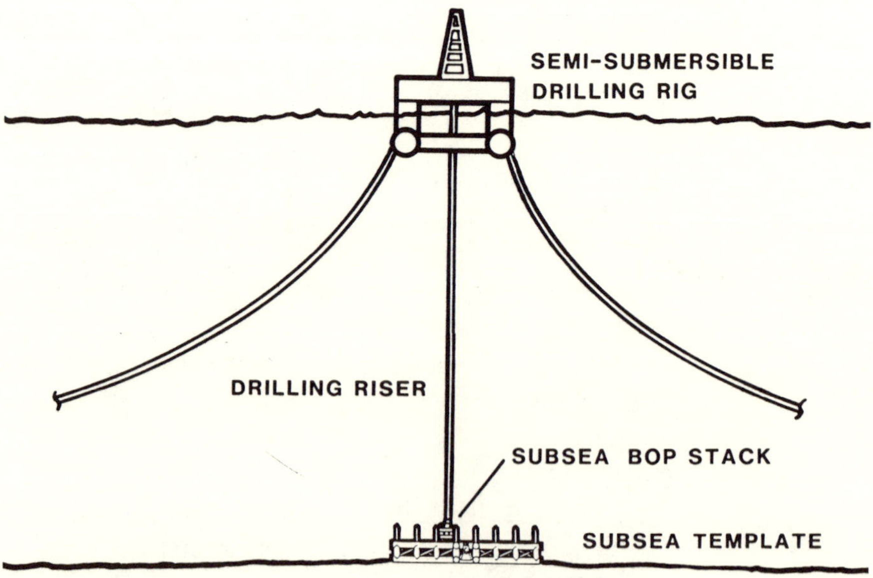

FIG. 4–3 Drilling from a floater. *(Courtesy Vetco)*

WELLHEAD SYSTEM

The subsea wellhead system run by the drilling vessel prior to installing the TLP is an 18¾-in., 10,000-psi WP system (Fig. 4–4), modified to accommodate the tieback system. The wellhead housing is a standard extra-long type except that the landing ring has been moved down so more of the housing is exposed when it lands inside the 30-in. housing. The 18¾-in. × 13⅜-in. casing hanger inner profile is the same as a standard 16¾-in. × 13⅜-in. casing hanger, which allows the same running tool to be used for both the 18¾-in. floater-drilled and 16¾-in. TLP-drilled wells.

DRILLING RISER SYSTEM

Drilling with a surface BOP and high-pressure riser is a reversal of the standard techniques used in drilling from a floating drilling vessel. With this arrangement, a smaller, less complex BOP at the surface can be maintained and operated more easily. The high-pressure drilling riser assembly is essentially an extension of the BOP, designed to withstand the maximum pressure and stresses anticipated during drilling of the well under all weather conditions (Fig. 4–5). The drilling-riser assembly consists of the lower riser package, the intermediate drilling riser joints, and the top section.

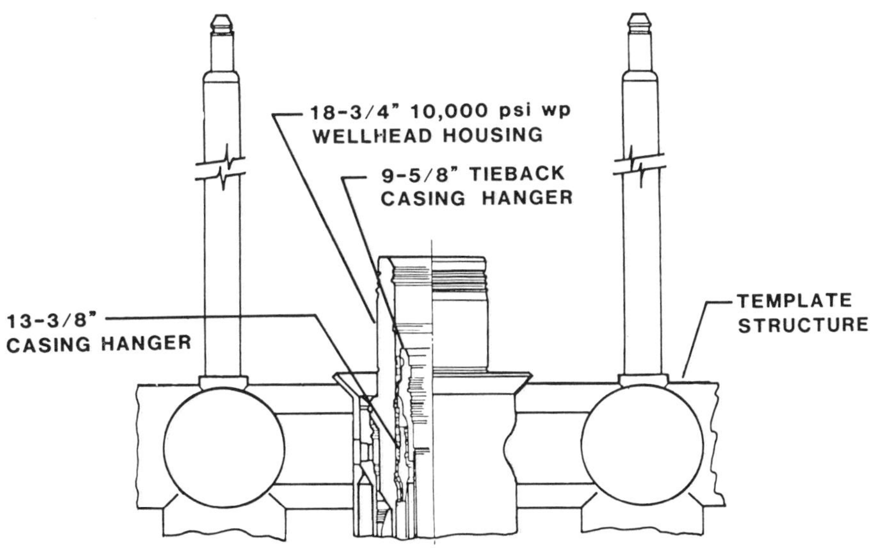

FIG. 4–4 18¾-in. wellhead stackup. *(Courtesy Vetco)*

The lower riser package is run at the bottom of the string and locks the riser to the subsea wellhead. It permits angular deflection of the riser with respect to the subsea wellhead due to surging of the TLP. The 16¾-in., 5,000-psi WP connector with metal-to-metal seals provides pressure integrity to the subsea wellhead. It will withstand riser tension and transmit bending moments without overstress or fatigue.

The 20-in. diameter riser string, extending from the lower riser package at the sea floor to the TLP substructure at the surface, is made up of 20-in. OD × 1.375-in. wall joints of 80 KSI minimum yield seamless pipe joined by high-pressure riser connectors. The integral connections are stronger in both tension and bending than the pipe body. Each joint contains three integral hydraulic lines for control of the connector in the lower riser package (lock, primary and secondary release). The riser joints are 50 ft long, and pup joints space out as required.

The top section of the drilling riser system includes the upper 16¾-in. diameter, 5,000-psi WP flex joint, drilling riser tensioner, tensioner spool, and

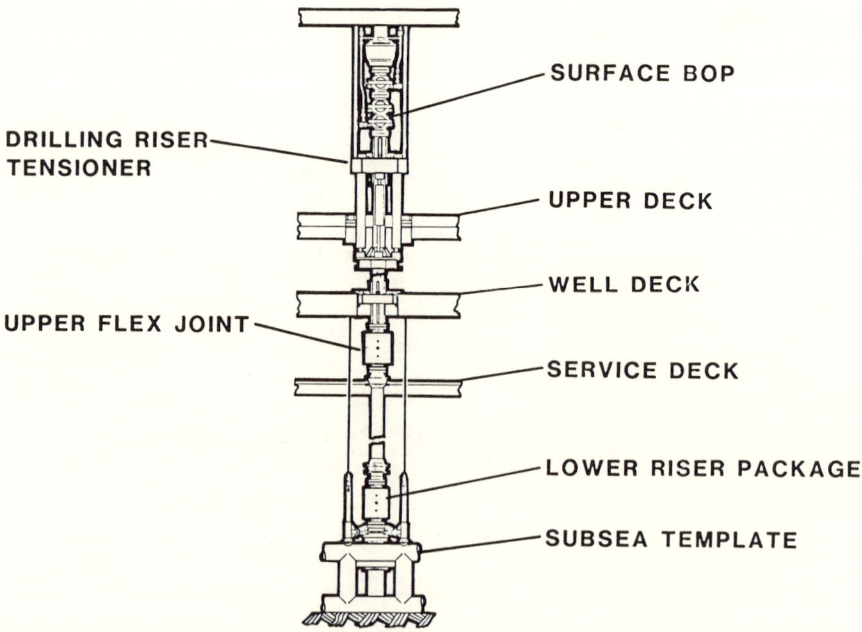

FIG. 4–5 High pressure drilling riser assembly. *(Courtesy Vetco)*

BOP stack. The upper flex joint is located just below the TLP deck levels and allows larger riser angles than could otherwise be accommodated within the TLP deck arrangements. Bearing points on the lower structural level of the TLP decks are required to maintain the upper portion of the drilling riser in a vertical position.

TENSIONING SYSTEM

Maintaining proper tension in the drilling riser is important to prevent buckling and to ensure a long fatigue life. Riser deflections due to winds, waves, currents, and platform offset require that the tensioning system allow changes in the apparent length of the riser relative to the TLP decks. The tensioner system supports the surface BOP from a point below while providing the 1,000,000-lb working tension anticipated for the high-pressure drilling riser. The tensioner module is a hydraulic-pneumatic system consisting of support beam structure, hydraulic cylinder, accumulators, air bottles, interconnecting piping, control console, air supply, and compressors with dryers (Fig. 4–6).

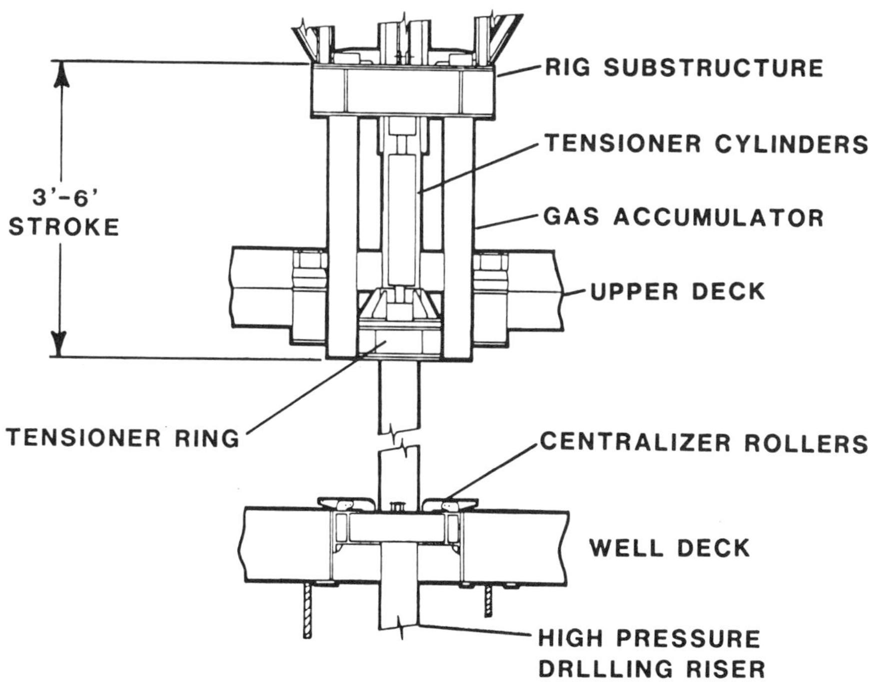

FIG. 4–6 Drilling riser tensioner. *(Courtesy Vetco)*

The drilling riser tensioner is designed to allow passage of the lower wellhead connector and high pressure flex joints so the riser may be run through the drilling riser tensioner, permitting a straightforward, simple running procedure.

MOORING SYSTEM

The TLP mooring system (Fig. 4–7) consists of the following:

- *Mooring templates are provided for each corner of the TLP structure. Each template carries landing receptacles for anchor piles, tension-leg assemblies, and a central running string. Each template is equipped with a template leveling system.*
- *An anchor connector/lower flex assembly is located at the bottom of each tension leg. This assembly provides an automatic mechanical latch to secure each tension leg to the mooring template. The built-in flex element allows angular movement of each tension leg relative to the latch mechanism.*
- *Tension leg joints are run and latched at each corner of the TLP structure and consist of heavy tubulars with a mechanical latch at the bottom. After latching onto the mooring template, the legs are tensioned adequately to pull the TLP down into its final position.*
- *A crossload bearing/upper flex assembly at the top of each tension leg is located in the individual mooring tunnel area. This flex assembly allows angular movement of the tension leg below the TLP and also takes up any side loads imposed by TLP movements.*

Along with the fact that foundations and risers comprise only a small percentage of total costs, the extensive use of standard equipment and techniques gives the tension-leg platform the ability to operate in water depths well beyond those possible with existing technology. The significant cost advantage is apparent. Cost reductions may also accrue by reducing the number of tubulars required for each well. The TLP concept may well provide economical feasibility and impetus for development of deep-water marginal fields.

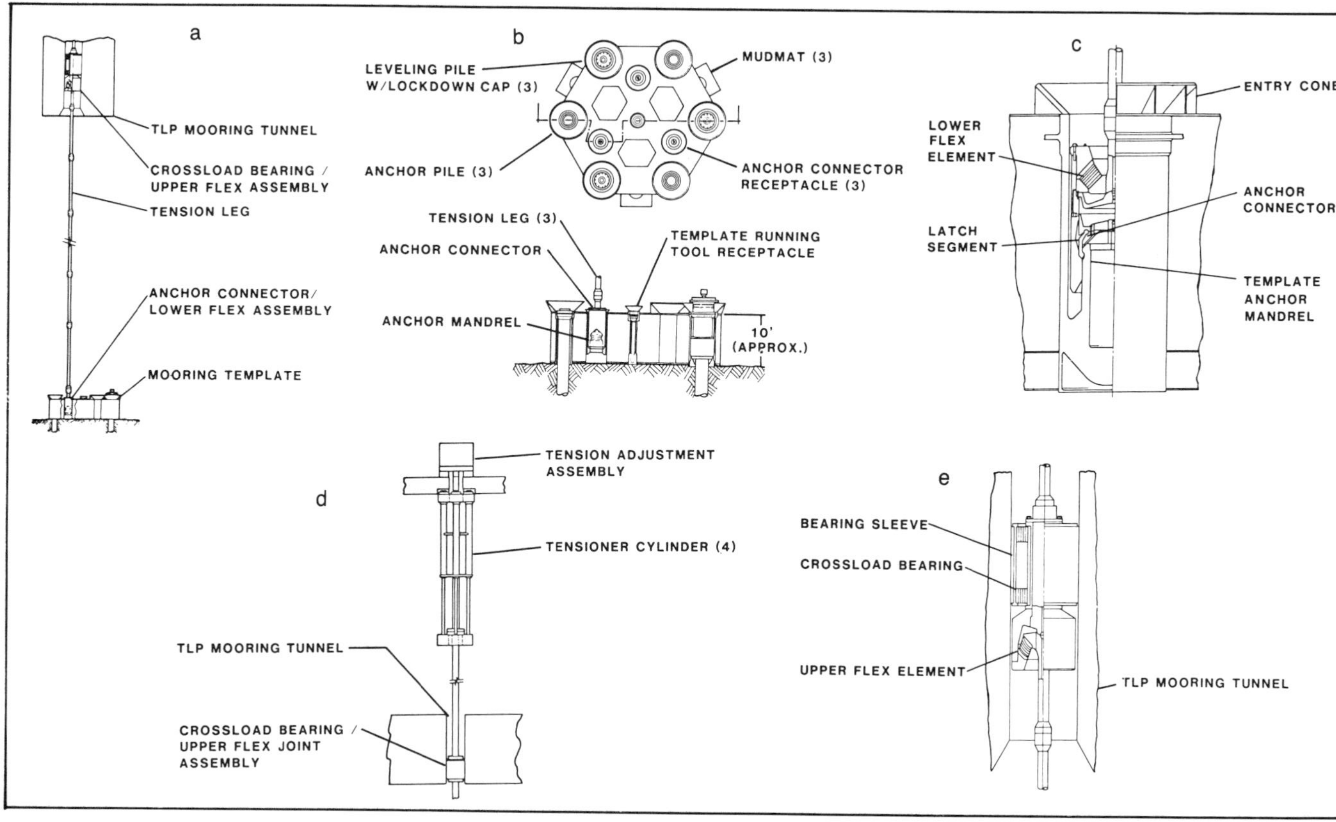

FIG. 4–7 TLP mooring system overview. *(Courtesy Vetco)*

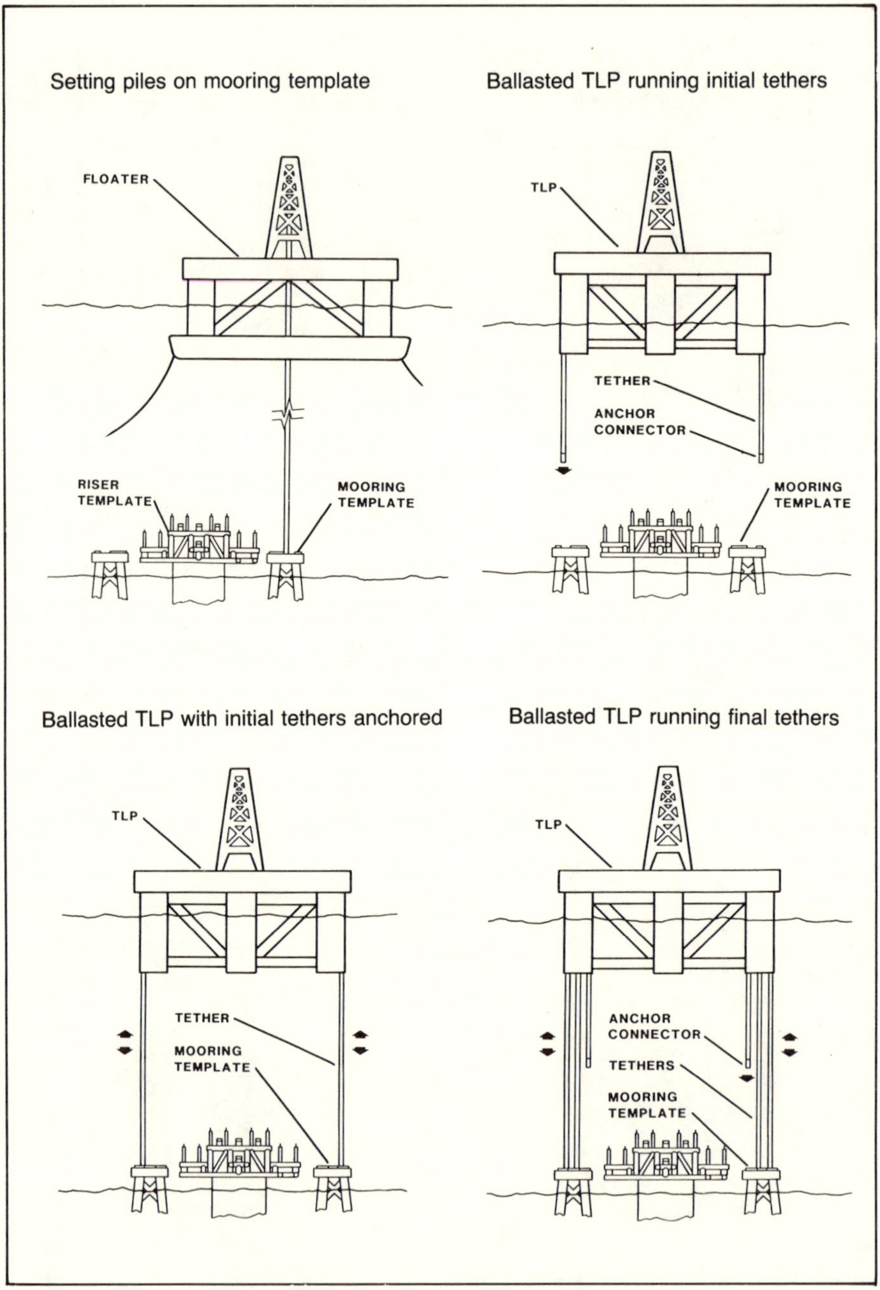

FIG. 4–8 TLP mooring sequence.

5 Template Utilization

The design, manufacture, and installation of subsea drilling and production systems utilizing templates has evolved into a successful, widely accepted technique. Template technology has proven its versatility in multiwell development from the North Sea to offshore Brazil.

After committing to development of a field using a clustered-well technique, one of the first decisions is which type of template is best suited for that application—spacer, modular, or unitized. The usual criteria for template selection are water depth, reservoir potential and configuration, number of wells, type of production system, and method of installation.

Verticality of the wells is, at best, difficult to achieve in any multiple subsea well system. As water depth and number of wells drilled increase, the problem is magnified. Consequently, the best assurance of verticality must be a prime consideration when deciding on the preferred type of template for a particular application. For instance, due to their more sophisticated design, modular and unitized templates provide better verticality and more precise alignment between the template and platform than the spacer template. For this reason, spacer templates are usually recommended only for small systems in shallow water for tieback to a platform.

SPACER TEMPLATE

As the name implies, the spacer template is of simple construction and provides wellhead receptacles on a known well-bore spacing. Since it has no built-in leveling system, there is no way to be sure each wellhead will have the necessary degree of verticality for tieback of a large number of wells in deep water. With only a gimballed permanent guide base to level each well, the spacer template also requires that the sea floor have a slope of less than 5° for efficient use. As a result, spacer templates are not recommended for tieback of more than six wells nor water depths greater than 200 ft. The first six-well spacer template was installed in 150 ft of water in the Kerindingan Field offshore Indonesia in 1973.

While spacer templates are usually larger than a drilling vessel moonpool (and must be installed by crane barge or by keelhauling), Fig. 5–1 illustrates a spacer template whose overall dimensions will let it be lowered through most moonpools or spider deck openings. It has four well slots and two guide pile slots. The wellhead receptacles are on a minimum 6- to 7-ft spacing to accommodate the retrievable permanent guide structure. This permits the use of standard guidance systems and blowout preventer stacks during drilling operations. Also, mud-line suspension equipment is compatible with spacer template design. For appropriate applications, such as jackup drilling, the simple spacer template affords much greater economy than more complicated template designs.

MODULAR TEMPLATE

The modular template is comparatively small, being made up of several interlocking modules. This type of template is particularly well suited for deep-water applications since it is designed specifically to be run on drill pipe through the moonpool of a floating drilling vessel. Modular templates are recommended for use in water over 200 ft deep with a relatively small number of wells. Although running procedures are similar to those of a standard guide base, the modular template is limited as to number of wells due to the requirement for a hangoff receptacle for each cantilevered well module.

The basic component of the modular template system is the base structure, much like a large temporary guide base, which supports the main template module. The support structure has a large bearing surface area and is equipped with a level indicator. A TV camera attached to the drill pipe running string provides monitoring capability for leveling the main template module. Level readings from the base structure provide the measurements for adjusting the two screw jacks on the main template module before it is lowered and landed. Precise leveling is then accomplished by further adjustment of the screw jacks after landing.

Fig. 5–2 illustrates a main template module with receptacles for four cantilever modules. This structure is used in place of a standard permanent guide base when drilling the first well of the modular template system. The main template module is run with the 30-in. diameter conductor string. Each

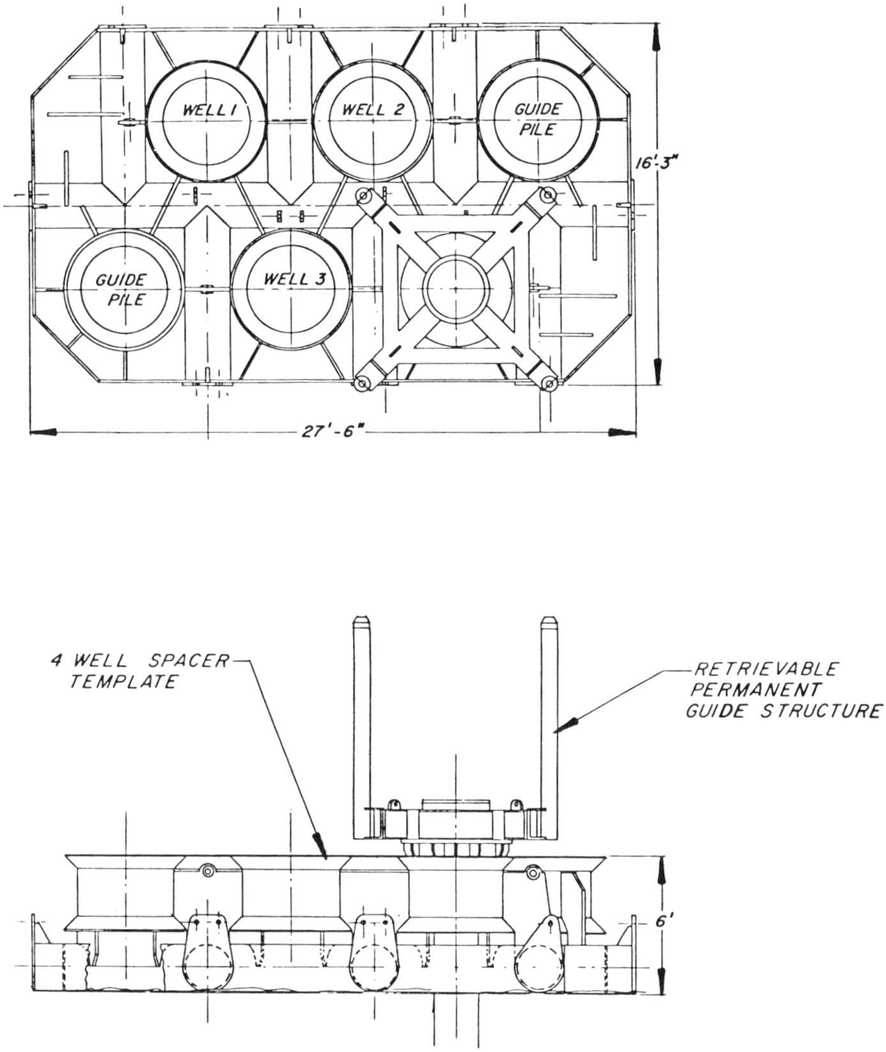

FIG. 5–1 4-well spacer template. *(Courtesy Vetco)*

hangoff preparation on the main structure consists of two vertical slots fabricated from heavy plate. Large-diameter horizontal pins are installed at the lower end of each slot and receive the lower hook assembly of a cantilever well module.

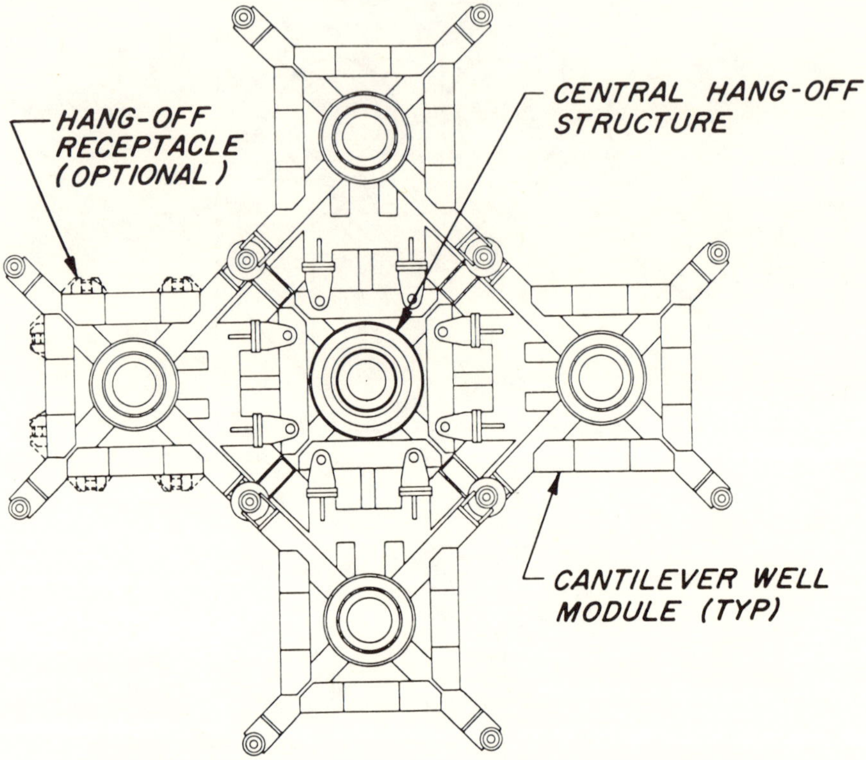

FIG. 5–2 5-well modular template plan. *(Courtesy Vetco)*

The cantilever well modules are run on drill pipe using two guidelines of the main template module. The only operational difference in drilling through a cantilever module is that the 30-in. conductor is installed through the module and is held down by a retaining ring instead of being run with the structure. Single-well base structure modular templates were installed in the South Kavala Field, offshore Greece, and in the Tiffany Field, North Sea.

A different style modular template is shown in Fig. 5–3. This is a three-well base structure that is run on the central 30-in. conductor string. The gimbal on the 30-in. extension (below the three-well base) lands in a standard temporary guide base and provides a leveling effect. This base has cantilever hangoff points for an additional six wells and guide modules for two jacket bumper piles. Lockdown mandrels for attaching a manifold module for the template trees may be incorporated if desired. This design of modular template

FIG. 5–3 3-well base structure. *(Courtesy Vetco)*

was installed in the Thelma Field in the North Sea in 1980. A modular template with positions for six wells—four drilled through cantilever modules and two through the main template module—was installed in 240 ft of water in the Fulmar Field, North Sea.

UNITIZED TEMPLATE

Unitized templates can be used with many types of production systems but are particularly suitable in fields requiring a large number of wells for development. Since the unitized type of template is made from large tubulars (30-in. diameter or larger) and in one-piece construction, it can be designed to handle the installation, leveling, and wellhead requirements for more wells than any other model (Fig. 5–4). The unitized template offers extreme water depth capability and, currently, up to 32-well capacity. In addition to superior structural integrity, the larger tubular members supply buoyancy to the template, which aids in installation.

The unitized template incorporates a heavy-duty three- or four-point leveling system. It may be lowered to the ocean floor by crane barge or by a drilling vessel's draw works. Integral or retrievable permanent guide bases adjust well spacing to meet platform slot requirements, thus affording maximum economy in platform and template.

Unitized templates have been installed successfully (with drilling or production currently in progress) in numerous offshore fields around the world: 15 wells, 450 ft of water, Garoupa Field, offshore Brazil; 20 wells, 480 ft of water, Northwest Hutton Field, North Sea; and 12 wells, 300 ft of water, South Montrose Field, North Sea. In 1980, a unitized template for producing ten subsea wells through a subsea manifold to a floating vessel was installed in 624 ft of water in the Enchova Field, offshore Brazil.

INSTALLATION

Installing any of these templates requires planning and coordination. It usually involves work boats, a barge or crane barge, and/or a floating drilling vessel plus acoustic or electric positioning equipment. Consequently, it is important to have properly trained, skilled personnel in control of the operation to ensure the template is landed and oriented properly without damage.

The size of the unitized template dictates that its installation normally would be the most complex. Unitized templates may be installed by a crane barge (Fig. 5–5), by partially submerging a conventional barge using the inherent buoyancy of the template to float it off the barge, or by slinging the template under a semisubmersible drilling rig and utilizing the rig lifting equipment for installation. When the template is placed in the water, buoyancy

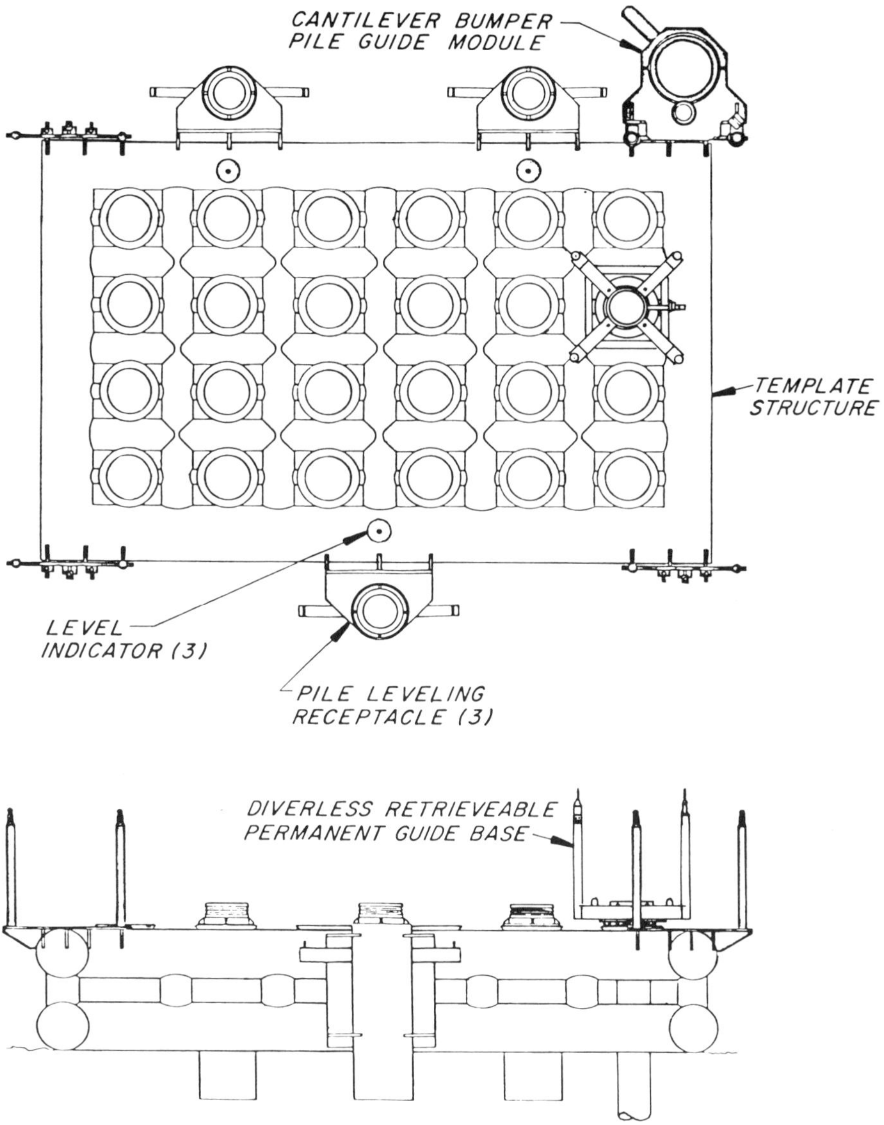

FIG. 5–4 24-well unitized template with diverless retrievable permanent base. *(Courtesy Vetco)*

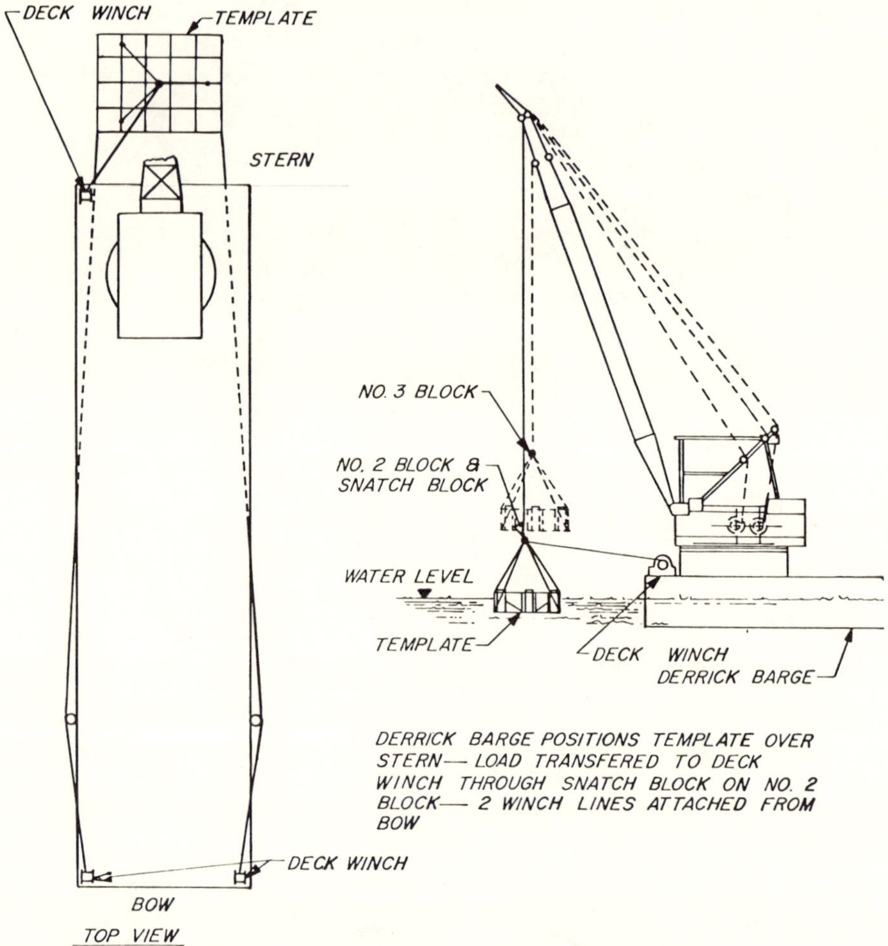

FIG. 5–5 Unitized template installation. *(Courtesy Vetco)*

can be maintained in the main structural section to facilitate controlled lowering. Once the template is floating, the sling supports can be replaced with lighter ones by a shallow diving operation. In addition, the lowering line can be supported at another block farther out on the crane boom, since the apparent weight of the template in the water is less than its weight in air due to its inherent buoyancy.

Installation using a semisubmersible is accomplished by transporting the template by barge to a sheltered area where the drilling semisubmersible is located. The cargo barge is brought underneath the semi, between the hulls, and directly below the derrick where lines are attached from the template to the draw works of the drilling vessel. The template is raised against the cellar deck floor and secured in preparation for the journey to the drilling site. After anchoring on location, the semi uses drill pipe and its draw works to lower the template to the ocean floor.

LEVELING

After the template is landed, the leveling system can bring the template from an initial inclination of up to 5° to a final position within 0.5° of horizontal. A level template is critical in order to reduce wear during drilling, to minimize bending moments and forces transferred to the template during drilling, and to facilitate engaging subsea stabs. Because a level template is so essential (particularly in a unitized template), a solid base must be provided so the template cannot shift and the leveling operation must be isolated from the floating vessel's erratic motions caused by sea forces.

The equipment that meets these rigid requirements uses 30-in. OD drilled and grouted piles installed through heavy-duty, one-way slips to ensure the solid base and to hold the template in position. After the template has been lowered and guidelines have been installed, the drilling assembly is run through the pile guide housing and the hole is drilled for the pile section.

After the template has landed, the large hydraulic leveling tool is latched to one corner of the template. The hydraulic fluid pumped to the leveling tool from a small surface pump strokes the leveling cylinder against the pile smoothly and accurately, lifting that corner of the template to the desired position (Fig. 5–6).

To ensure trouble-free leveling of the template and drilling of template wells, a centralizer should be used when drilling hole for the leveling piles, 30-in. × 20-in. × 13⅜-in. × 9⅝-in. × 7-in. diameter offshore exploration be installed prior to running the 13⅜-in., 9⅝-in., or 7-in. riser. Any of the located directly in line with the center of the receptacle. If the drill bit or hole opener is not centralized, the drill pipe may "walk" over against the receptacle wall, making the hole off center and increasing the difficulty of proper pile or conductor installation.

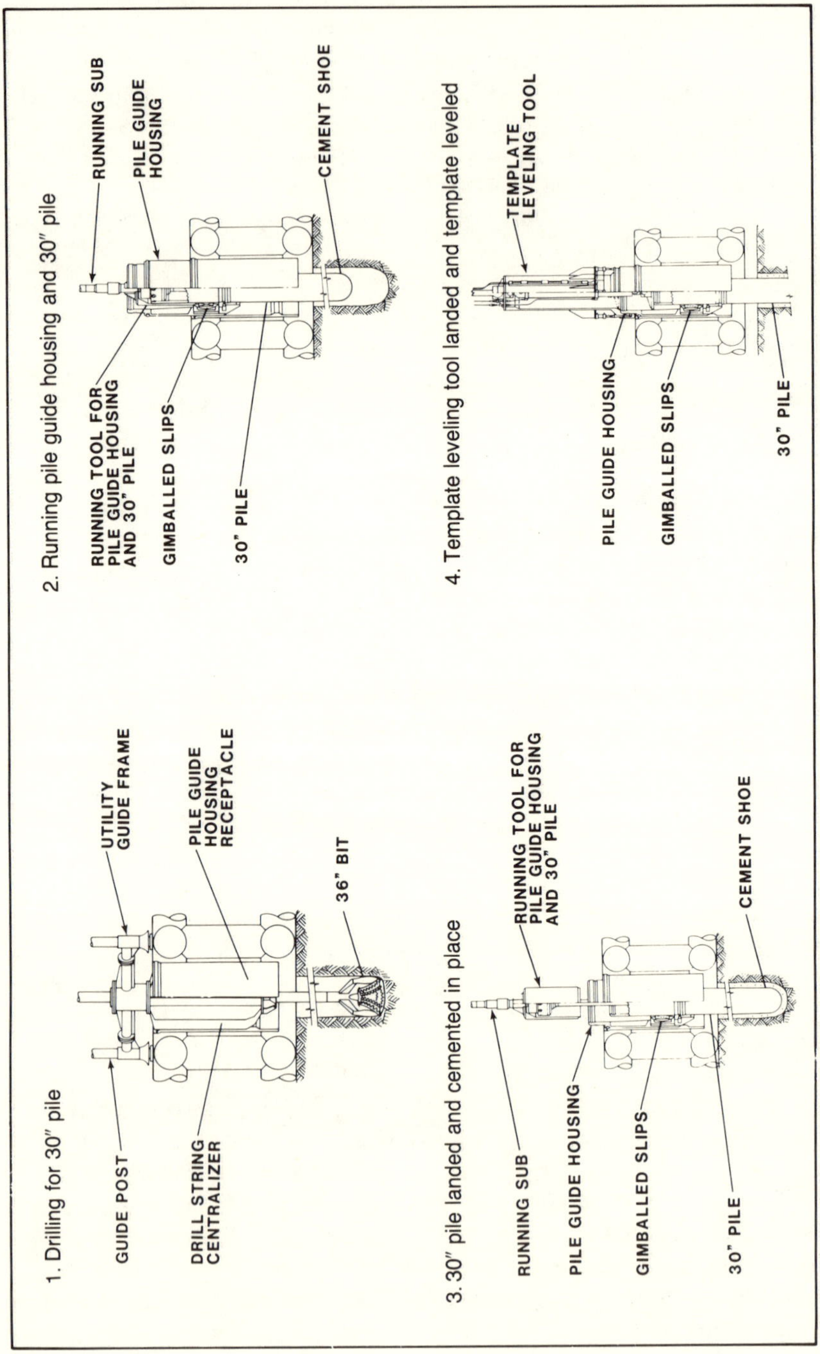

FIG. 5–6 Template-leveling operation using 30-in. piles. *(Courtesy Vetco)*

6

Mud Line Casing Suspension and Casing Support Systems

Mud line casing suspension and casing support systems are designed specifically for drilling from a bottom-supported fixed or movable platform. The mud line suspension system utilizes running and tieback tools for installation, removal, and reinstallation of the casing strings, permitting temporary abandonment of the well after drilling from a mobile platform. The mud line casing support system normally is used from a permanent platform installation and is made up directly to the casing running string without running or tieback tools. Both systems are of simple design with smooth OD profiles for easy installation in minimum vessel time.

A mud line suspension system (Fig. 6–1) provides a means of hanging off the various casing strings at or below the mud line while providing a method of disconnecting all casing strings. In addition, this method of supporting the casing strings reduces the load carried by the drilling platform since the weight of the casing strings is supported at the mud line. The selected concentric casing strings extend from the mud line to the drilling platform where conventional land-type blowout preventers are installed for pressure control during drilling operations.

CASING INSTALLATION

In most cases, a mud line suspension system will utilize the standard 30-in. × 20-in. × 13⅜-in. × 9⅝-in. × 7-in. diameter offshore exploration casing program. Initially, the outer conductor casing string (30-in.) is jetted, drilled, or driven into place, allowing the 30-in. landing ring to be located at any desired elevation with reference to the mud line. This conductor string

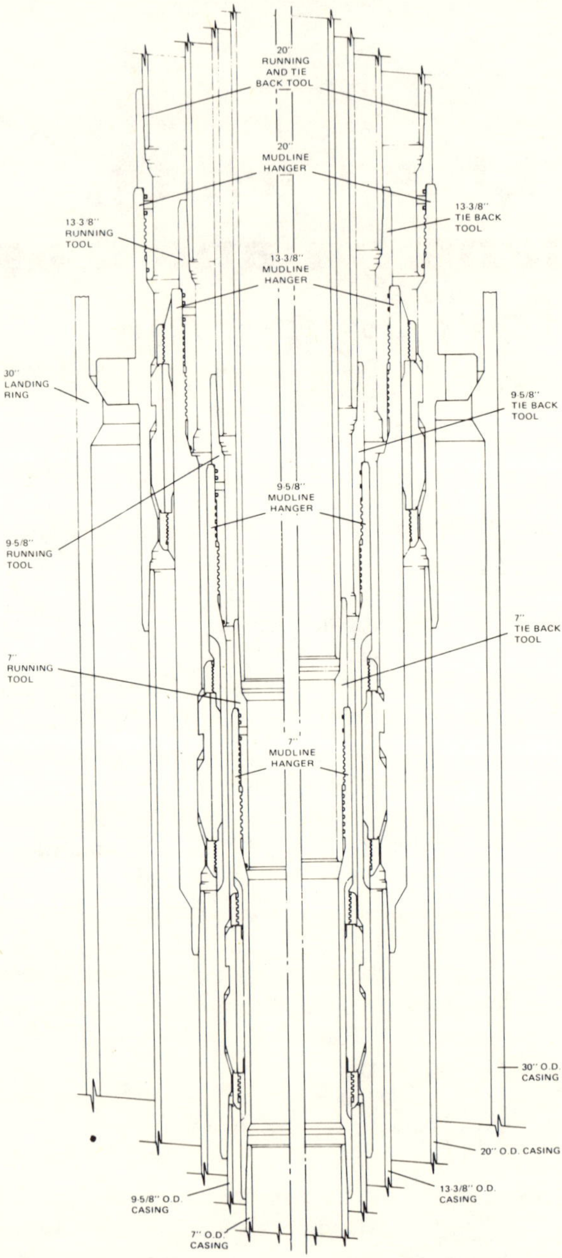

FIG. 6–1 Mud-line suspension with running and tieback tools. *(Courtesy Vetco)*

extends to the drilling structure where low-pressure flow diverter equipment is then installed for protection during the drilling, running, and cementing sequence of the next string of casing.

The 20-in. casing string is suspended on the 30-in. landing ring by the 20-in. mud line casing hanger and normally is extended back to the drilling platform. All subsequent casing strings are run and suspended in a similar manner, each string supported at or near the mud line by a mud line casing hanger of the appropriate size with threaded preparation for both the running tool and tieback tool.

TEMPORARY SUSPENSION OF WELL

Sometimes it is necessary to suspend the well after the drilling phase is completed and prior to installing a permanent production facility. The same running tool used to run each of the strings is utilized to strip off each string from its respective casing hanger. When the casing strings (from mud line to drilling platform) have been retrieved, a temporary well abandonment cap or protective cover is installed to reduce marine growth and corrosion in critical areas of the casing hangers (Fig. 6–2).

PRODUCTION TREE INSTALLATION

A well completed by the mud line suspension method may use a subsea production tree, making the well a true subsea completion. As an option, any and all of the casing strings can be extended back to a production platform on the surface by means of tieback assemblies for each of the casing strings. In such cases, a conventional land-type production tree is installed.

PLATFORM ECONOMICS

Since most of the weight of the casing strings is suspended at or near the mud line and is not transmitted to the platform structure, a production platform of more economical construction may be used with mud line suspension systems.

OPERATING PROCEDURE

The following method utilizes a mud line suspension system in a drilling program. In actual practice, however, the operator must carefully consider factors known only to him regarding the particular drilling area before implementing any procedure.

FIG. 6–2 20-in. mud-line corrosion cap and running tool. *(Courtesy Vetco)*

Before the 30-in. diameter conductor and housing may be run, bottom conditions must be tested. A bit run on drill pipe will afford an accurate measurement of water depth and will identify bottom conditions. At this time it is determined whether the 30-in. conductor should be jetted, driven, or run into a drilled hole and just where the 30-in. housing is to be landed with respect to the mud line. The conductor string is then spaced out to land the housing at the correct depth.

After the 30-in. conductor has been cemented in place, a 26-in. concentric hole is drilled to the desired depth and is conditioned. All relevant measurements are checked on the 20-in. casing string to ensure the last collar is clear of the 20-in. casing head. The running tool is made up to the 20-in. casing hanger and the 20-in. casing is run, slowing down when approaching the 20-in. casing seat. The hanger is landed and checked for correct seating. Then the 20-in. casing is cemented as required and any cement in the 30 × 20-in. annulus above the hanger is washed out.

To run the 13⅜-in. casing and hanger, a 17½-in. hole is drilled to the required depth and conditioned, and the drillstring is pulled out. Again, all relevant measurements are checked to be sure the last collar is clear of the casing head. The running tool is made up to the 13⅜-in. casing hanger and the casing is run. After the hanger is landed, measurements are checked for correct seating. When the lock ring of the 13⅜-in. hanger has engaged the locking groove of the 20-in. hanger, the 13⅜-in. casing is cemented as required with returns through ports on the 13⅜-in. hanger. Any cement rise in the 20-in. × 13⅜-in. annulus above the 13⅜-in. hanger is washed out through the ports on the running tool. These ports may be opened fully by backing out the running tool.

An alternative method after washing out is to back out the running tool, pull it, and run a tieback tool that makes up in the upper part of the 13⅜-in. hanger. The tieback tool (Fig. 6–3) has a metal-to-metal seal in the lower section. With either tool correctly seated, the seal is pressure tested and surface connections are installed.

The 9⅝-in. and 7-in. casing and casing hangers are run, cemented, washed out, and tested using the same procedure as for the 13⅜-in. Since the maximum OD of the 7-in. hanger is 8.523 in., 9⅝-in. casing of over 47 lb/ft weight should not be used in either the running or tieback string if a 7-in. hanger is to be run later.

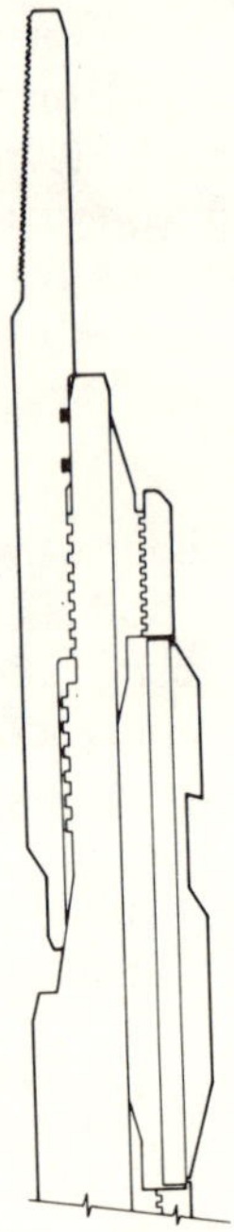

FIG. 6–3 Tieback tool.

Alternative latching systems are available for mud line suspension systems to meet specialized needs. Gray Tool Company's model DJEL (expanding latch) system offers a choice to operators wishing to land a mud line suspension latch-type hanger by downward movement of the casing only. The DJEL is suitable for all types of drilling operations, with special capabilities for production and directional drilling. It is best suited for situations in which reciprocating motion is undesirable, such as deviated drilling, tight holes, and loose formations.

It is unnecessary to pick up the string to set the DJEL hanger because the expanding latch segments automatically engage the recess in the outer housing as the casing is run in. The segments are then locked in place and can be released only by picking up the casing string.

To cap the well for abandonment or subsequent reentry, plugs are set in the 7-in. casing as required. If the running tools have been left in position on all the casing strings made up to their respective hangers, each string is then recovered by releasing the running tool with right-hand rotation, pulling the string, and laying down. If tieback tools have been left in position on the casing strings, the tools are released with left-hand rotation and the strings are pulled and laid down. The 20-in. running string is always recovered with a right-hand rotation, pulling and laying down.

The 20-in. left-hand thread corrosion cap is run inside the 30-in. conductor string using the J-slot running tool on drill pipe. The corrosion cap is made up to the 20-in. mud line casing hanger. The running tool is released from the corrosion cap by setting down, rotating to the right, and picking up the string. The 30-in. conductor string can be retrieved by having a diver release the locking ring on the 30-in. connector at the mud line. If a hydraulic connector has been used, the release may be accomplished from the surface by applying sufficient operating pressure.

REENTRY INTO A CAPPED WELL

This discussion assumes that a fixed platform is being used and that it is positioned correctly over the well. The 20-in. corrosion cap J-slot running and retrieving tool is run and diver-stabbed into the 20-in. corrosion cap. (A surface valve should be installed on the retrieving string in case wellhead pressure is released by the stinger on the retrieving tool.) Right-hand rotation releases the corrosion cap so it may be picked up and pulled to the surface. The 20-in. casing riser, with the 20-in. mud line hanger running and tieback tool on bottom, can now be run and diver-stabbed into the 20-in. mud line hanger. When the seal is complete, it may be pressure tested.

Depending on requirements, the surface wellhead and BOP stack may be installed prior to running the 13⅜-in., 9⅝-in., or 7-in. riser. Any of the risers can now be run (with the choice of the left-hand thread running tool or the right-hand thread tieback tool) to make up and seal the desired mud line hanger. For permanent completions, the tieback tool would be used because of its right-hand thread makeup and the metal-to-metal seal area in its lower section. After pressure testing is successfully performed, the completion program may proceed. It is suggested that a "sugar pill" or commercial cement retardant be spotted through the running string at the casing hanger following washout on all strings in case of cement expansion in the annulus as the cement sets.

III

Completion Systems

Mary Lou Moore *Vetco Offshore Inc.*
Charles D. Bridges *Gray Tool Co.*

7 Platform Completions

Drilling and establishing a well requires placing equipment from which the various strings of downhole casing are suspended. In a land completion, large-diameter pipe (frequently 30-in.) is placed or driven into the ground deep enough to provide a stable environment through which drilling operations can be performed. This *conductor pipe* extends only slightly above ground level and supports the weight of all subsequent equipment—heads, spools, hangers, packoffs, and eventually a Christmas tree. In adapting land procedures to shallow-water operations (generally 50-ft water depths or less), all that is necessary is to have the conductor pipe extend to the level of the platform deck. The heads, spools, and hangers that suspend downhole casing and seal off well pressure are supported by the conductor pipe. Although the platform itself does not support the weight of the wellhead, it must be large enough and strong enough to hold all of the equipment necessary for the drilling and completion operations.

A shallow-water completion is similar to a land completion, but there are variations in the equipment required. For safety and environmental protection, a valve must be installed in the tubing (production) string below the mud line (sea floor). This downhole safety valve (often called a subsurface safety valve or SSSV) is a hydraulically operated fail-safe ball valve or flapper valve that will seal off the flow through the production string in an emergency. So the SSSV can be operated from the platform, control lines must lead to it, necessitating equipment with control-line passages. If the control lines go through the tubing hanger, the passages may be vertical and extend completely through the hanger with the control line then passing through either the flange or the side of the tubing bonnet (Fig. 7–1). The passages also may be L-shaped within the hanger with the control lines passing through either the flange or the sidewall of the tubing head. The latter arrangement allows one to install and test the control lines prior to removing the BOP stack and installing the tree—an advantage in time and safety.

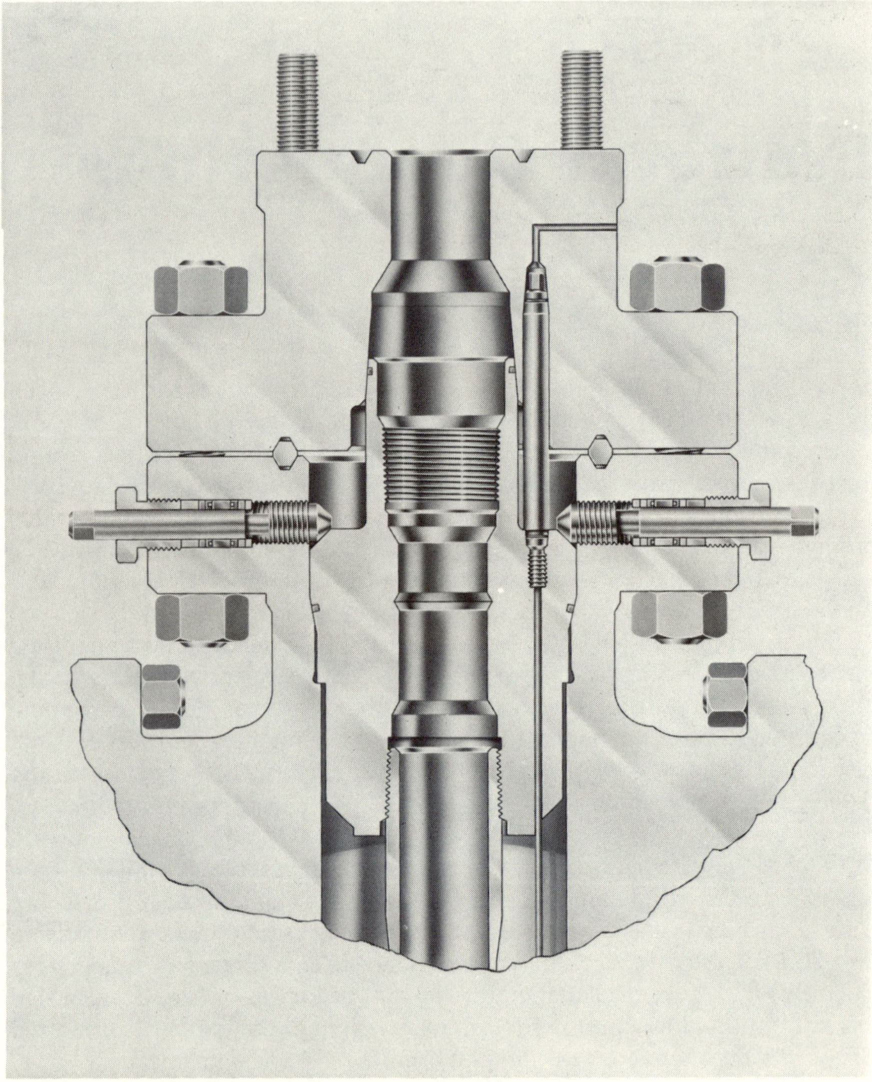

FIG. 7–1 Example of a Gray tubing bonnet with a control line outlet through the side.

There also may be a need for a downhole submersible pump if bottom-hole pressure is insufficient for economical flow or if the viscosity of the oil is too great for flow. Again, there must be a passage for the electrical cable that provides power to the pump, and any components through which the cable passes must provide a positive seal around that cable. When the production tubing is run downhole with the submersible pump and SSSV, the SSSV control line and the pump's electrical cable lead back to the platform. When the tubing hanger is installed to suspend the tubing, the cable and the control line are connected as required (Fig. 7–2).

The higher cost of offshore operations, as compared to land operations, has led to the development of several time- and space-saving devices. Notable among them is the compact multibowl wellhead (Fig. 7–3). Some suspend as many as four casing or tubing strings, eliminating several steps in the drilling operation. The BOP stack can remain in place during drilling and running the casing strings rather than being connected and disconnected several times. This provides better control of well pressure and safer working conditions. The multibowl head also eliminates the need for using a different head or spool along with flanged or clamped connections for each casing string, thus requiring less vertical space on the platform. However, this kind of wellhead is not suitable for all situations. It is best suited for a well-defined field in which formation depths are known precisely and where there is no expectation of difficulties.

It is common practice, especially in deeper water, to drill and then shut in a group of development wells from a mobile drilling structure such as a jackup or a semisubmersible rig that then moves on to another location. The completion platform is moved onsite at a later time to reopen the wells for production. This approach often saves time by permitting drilling to be done while the platform is under construction in the shipyard and frees expensive drilling equipment for other jobs. As the well is drilled, the strings of downhole casing (typically 20-in. 13⅜-in., 9⅝-in., and 7-in. diameter) are run and suspended at the mud line. After all drilling is completed, the well is plugged and capped.

The mud line suspension equipment must include tieback devices to which the abandonment caps can be attached (Fig. 7–4). The tieback assembly must have circulation ports so the areas above the hangers and around the tieback threads can be flushed clean of cement and debris. The number of caps may vary. The operator may install a cap for each casing string, for the 7-in. casing only (with an exterior shield to protect the rest of the system), or for both the 7-in. and 20-in. casing strings. If more than one cap is installed, the smallest is installed first and the largest is last.

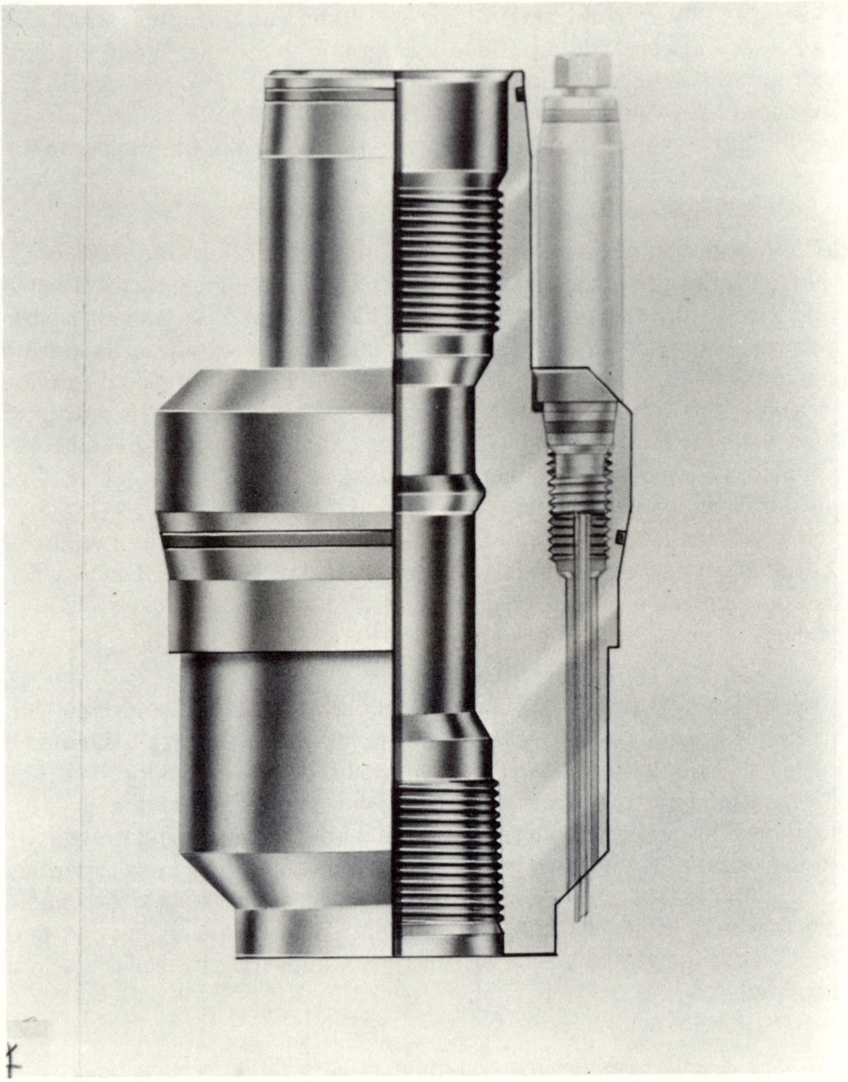

FIG. 7–2 Mandrel tubing hanger with control line arrangement.

Once the permanent production platform is in place, the wells must be tied back to it with casing extensions, called *risers*. The number of strings tied back varies according to preference and conditions, but usually the 20-in., 13⅜-in., and 9⅝-in. diameter strings are tied back. The precise procedure

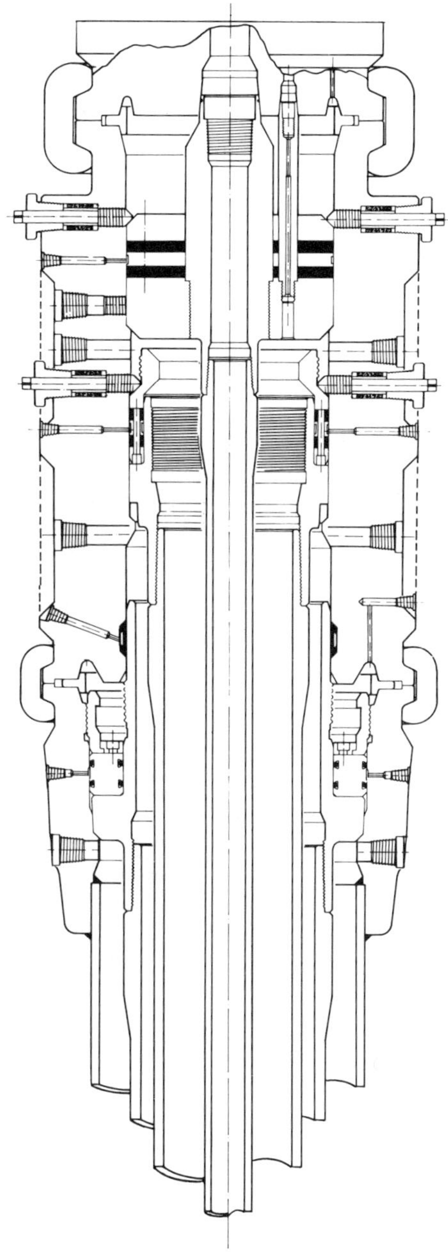

FIG. 7–3 Compact multibowl wellhead.

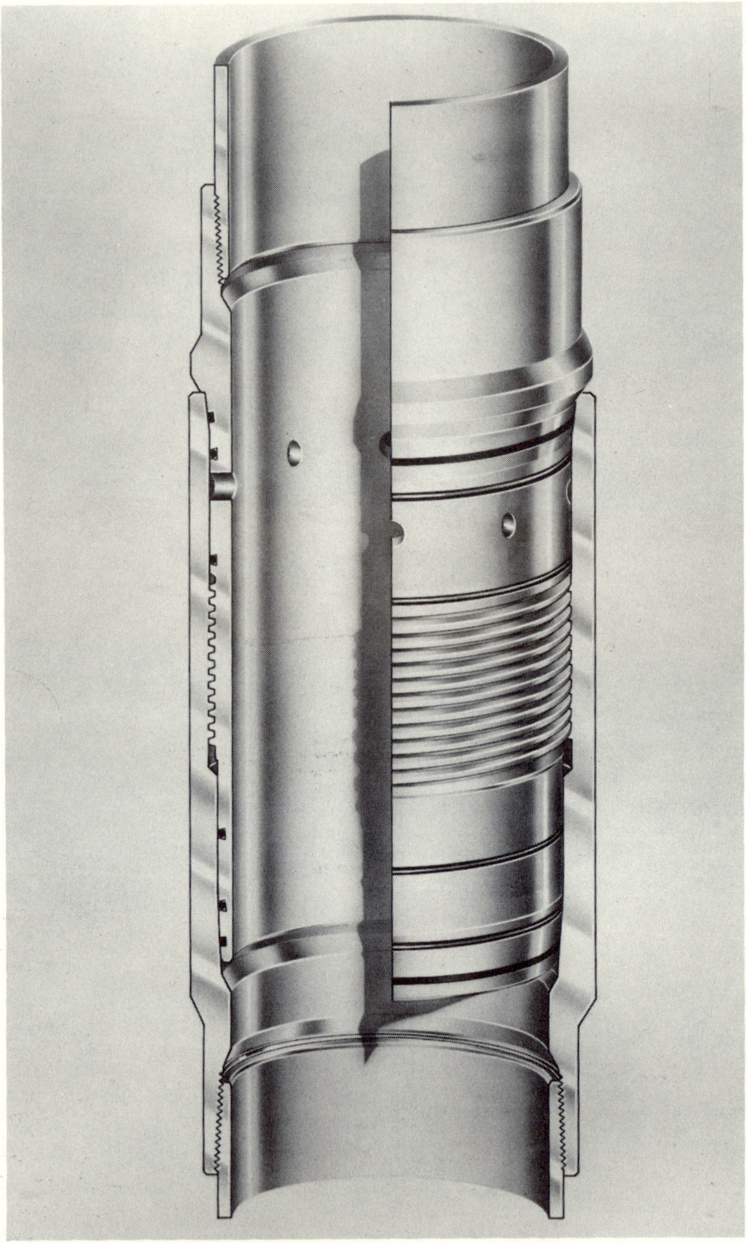

FIG. 7–4 Tieback seat and collar.

will vary according to the placement of abandonment caps, but the largest size casing string will be tied back first and the smallest size last. The risers connect to the same tieback assembly in which the caps were installed.

In shallow-water operations, the risers may be rotated to thread into the tieback assembly (Fig. 7–5). But in deeper, rougher waters, stab-in procedures and equipment may be necessary. Each riser string is hung off in a head at the platform level by pulling tension on the string, installing the casing hanger around the riser, and slacking off on the string to activate the hanger slips. The hanger usually has manually energized seals because, when the riser is slacked off, the weight may be insufficient to energize them. The riser pipe is essentially free-standing, its weight supported neither at the mud line nor at the platform. Buoyancy modules may be integral to the risers, assisting in keeping load off the wellhead. In a close grouping of wells, lateral support may also be provided for the riser strings. The tubing hanger is generally a mandrel type in order to provide access to control lines. The production tubing is hung off at platform level, and the Christmas tree is installed onto the tubing head.

By American Petroleum Institute (API) definition, a Christmas tree is an assembly of valves and fittings used for production control that includes all equipment down to the tubing head top connection (Fig. 7–6). The flow from the well comes through the production tubing to the tree and from the tree to processing or storage.

The tree is the primary point at which the flow can be controlled. Basically, the tree is a series of valves, some manually operated and some either pneumatically or hydraulically operated. The valve closest to the tubing head is called a master valve. A tree may have as many as three master valves arranged consecutively in the flow stream.

The topmost valve in the vertical run of the tree is called a swab valve or crown valve. The swab valve provides vertical access to the well for any necessary downhole work. Installed on the side of the tree is the wing valve, the last valve through which flow passes. Often, a choke is installed downstream from the wing valve for additional flow volume control. Should there be reason to halt the flow from the well, the choke is closed first. Then the valves are closed, starting with the wing valve, progressing in sequence to the lower master, and then to the SSSV.

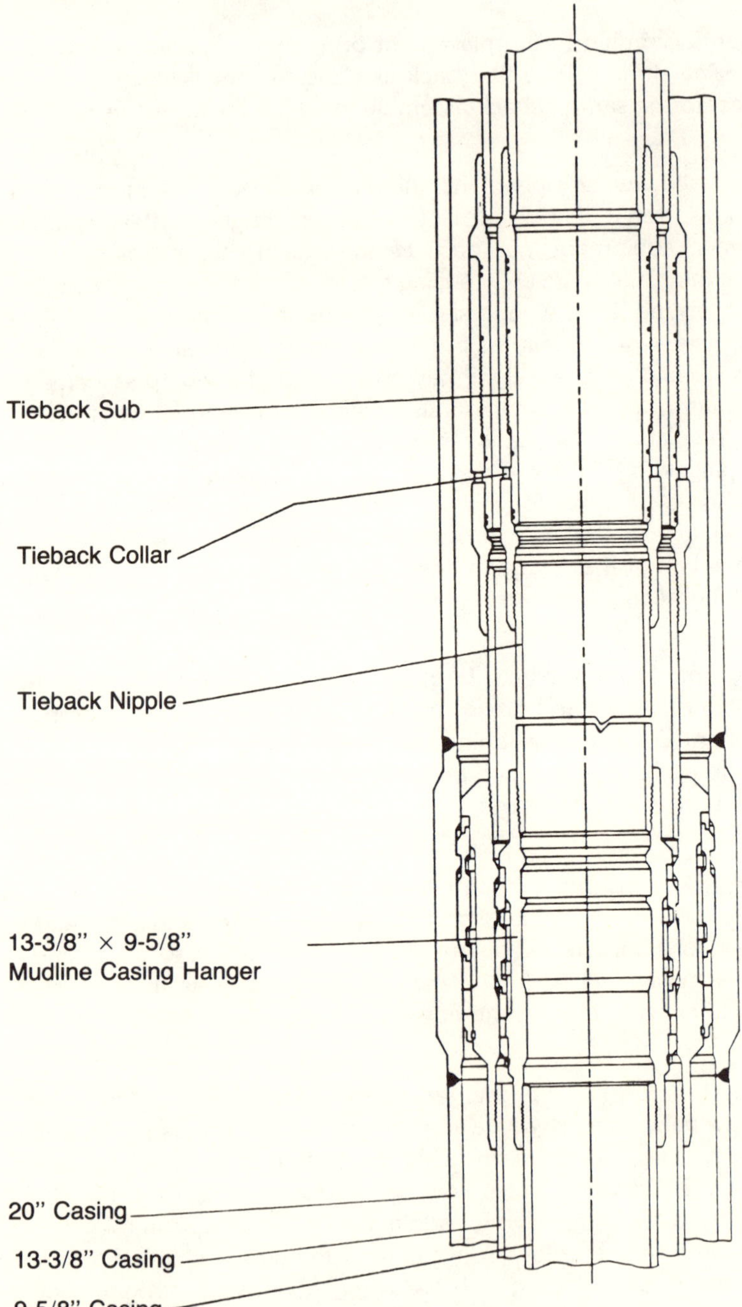

FIG. 7–5 Threading a riser into a tieback assembly.

FIG. 7–6 Christmas tree.

The size of trees varies considerably, according to well pressure and the bore size of the completion equipment. Trees range in height from 6 to 16 ft. The valves and other equipment may require a lateral space of about 10 ft. In a land completion there is usually no concern over the space occupied by a tree; but on a platform, space is a primary consideration.

DRILLING AND PRODUCING OFFSHORE

As noted earlier, offshore platforms today usually serve many completions, receiving flow from perhaps dozens of wells. Using ordinary trees would require a platform with much wasted deck space between trees. In response to the need for better space utilization, the industry has developed extremely compact trees for offshore use. Unitized, solid-block trees contain the necessary valves and require a much smaller area.

For use on a platform in the Santa Barbara Channel, Gray Tool Company designed such a tree with a height of only 32 in., including the hub for a clamp connection. Maximum horizontal space required for this tree is only 48 in. This small size was attained despite the fact that the tree contains five valves, one with an actuator. The platform also served as a drilling platform for the wells. It has 48 drilling slots with two drilling rigs that are skidded from one drilling slot to the next. Both drilling and production are in progress at the same time. With space at a premium, the use of unitized, solid block trees and clamp connections enabled substantial savings in platform size and construction costs.

Also true in a land completion, an offshore completion may have two or more production strings, terminating in producing formations at different depths. Multiple completions require the same equipment that single completions require (except for special tubing hangers that can suspend the necessary number of tubing strings) plus a Christmas tree with the required number of valves for each completion. Here, too, space requirements have led to the development of very compact trees and valve arrangements that occupy minimum space (Fig. 7–7).

Besides the SSSV described earlier, a wellhead surface safety valve (SSV) is required for offshore service. The API requires that this valve be second in the wellhead flow stream and provide positive well-stream shutoff if an emergency requires it. The valve must be of the normally closed type, which closes automatically in the event of loss of the hydraulic or pneumatic pressure that keeps it open. The API has also established standards for fire resistance of valves. The standards specify the maximum leakage permitted and the minimum degree of functioning required after a valve has been enveloped in a very intensive flame.

The technology of platform completions has evolved over the years in response to technical, financial, safety, and environmental demands. Certainly this evolution will continue as the need grows to drill in deeper waters.

FIG. 7–7 A 45° minidual valve.

8

Through-the-Leg Completions

Through-the-leg wells are drilled and completed inside a structural leg of an offshore platform. The basis and justification for this approach to platform well installation are discussed in Chapter 2.

Because the space allotted for the individual well's equipment in the multiwell clusters typical of through-the-leg installations is considerably less than in normal platform well installations, working space and access to the component parts of each well is restricted (Fig. 8–1). For this reason and because the tight configuration demands accuracy in placement of each component, it is essential to design through-the-leg projects as a complete system.

HARDWARE

In terms of completion hardware, through-the-leg wells utilize mostly standard internal components. Some modifications in the configuration of surface equipment are necessary or desirable to improve access both during installation and while in operation. As an example of this, dual valve body castings that place the valve stems in a narrow V position (instead of traditionally opposite each other) allow access for operation, maintenance, and repair from the outboard side of the circular well cluster. The internal components, actuators, and controls remain identical to a standard dual valve in opposing configuration (Table 8–1).

In fact, outlets on all wellhead components are arranged so full access to all valves is retained throughout the multiwell cluster. The positioning of valves, actuators and chokes on multiple dual-completion trees within a limited work area (such as in multiwell through-the-leg installations) requires specialized valve body configurations.

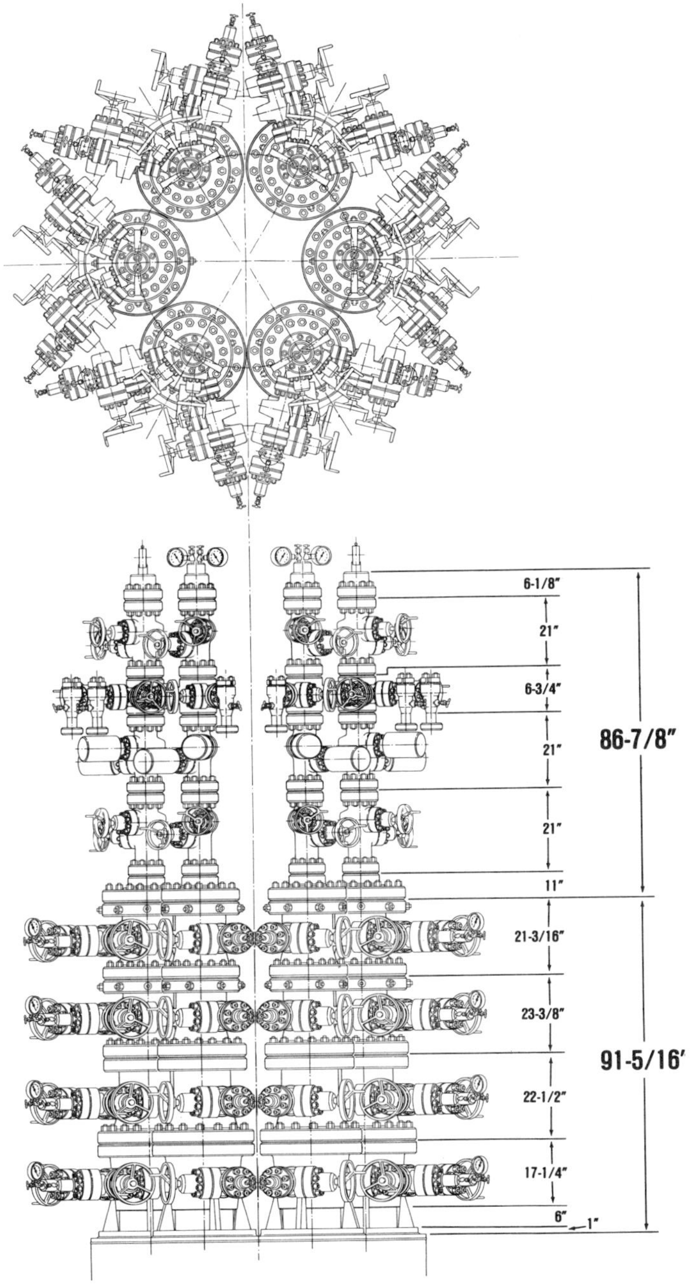

FIG. 8–1 Design of a through-the-leg completion system.

Table 8–1 Dual Completion Valves Specifications and Principal Dimensions

Dual Valves with D Flanged Ends

A Bore Size	Flanged	MSP (psi)	B ℄ to Top	D	F Over-all Length	G	R_1	R_2	℄	Number of Turns
$2\frac{1}{16} \times 2\frac{1}{16}$	D-Dual	3,000	$19\frac{7}{8}$		$12\frac{1}{2}$	21	1.773	1.773	$3\frac{35}{64}$	15
$2\frac{1}{16} \times 2\frac{1}{16}$	D-Dual	5,000	$19\frac{7}{8}$	$12\frac{1}{2}$	21	$2\frac{3}{16}$	1.773	1.773	$3\frac{35}{64}$	15
$2\frac{9}{16} \times 2\frac{9}{16}$	D-Dual	3,000	$20\frac{5}{8}$	$13\frac{3}{4}$	$25\frac{3}{4}$	$2\frac{1}{2}$	2.000	2.000	4	18
$2\frac{9}{16} \times 2\frac{9}{16}$	D-Dual	5,000	$20\frac{5}{8}$	$13\frac{3}{4}$	$25\frac{3}{4}$	$2\frac{1}{2}$	2.000	2.000	4	18
$3\frac{1}{16} \times 3\frac{1}{16}$	D-Dual	3,000	$23\frac{1}{2}$	$16\frac{1}{2}$	$26\frac{3}{8}$	$2\frac{1}{2}$	2.523	2.523	$5\frac{3}{64}$	$22\frac{1}{2}$
$3\frac{1}{16} \times 3\frac{1}{16}$	D-Dual	5,000	$23\frac{1}{2}$	$16\frac{1}{2}$	$26\frac{3}{8}$	$2\frac{1}{2}$	2.523	2.523	$5\frac{3}{64}$	$22\frac{1}{2}$

Mini-Dual Valves with GRAYLOC® Ends

A Bore Size	GRAY-LOC Hub	MSP (psi)	B ℄ to Top	C Wheel to Wheel	F Over-all Length	R_1	R_2	℄	Number of Turns
$2\frac{1}{16} \times 2\frac{1}{16}$	F × F	3.000	$19\frac{7}{8}$	$39\frac{3}{4}$	17	1.773	1.773	$3\frac{35}{64}$	15
$2\frac{1}{16} \times 2\frac{1}{16}$	F × F	5.000	$19\frac{7}{8}$	$39\frac{3}{4}$	17	1.773	1.773	$3\frac{35}{64}$	15
$2\frac{9}{16} \times 2\frac{9}{16}$	G × G	3.000	$20\frac{5}{8}$	$41\frac{1}{4}$	$18\frac{1}{4}$	2.000	2.000	4	18
$2\frac{9}{16} \times 2\frac{9}{16}$	G × G	5.000	$20\frac{5}{8}$	$41\frac{1}{4}$	$18\frac{1}{4}$	2.000	2.000	4	15
$3\frac{1}{16} \times 3\frac{1}{16}$	H × H	3.000	$23\frac{1}{2}$	47	$22\frac{3}{4}$	2.523	2.523	$5\frac{3}{64}$	$22\frac{1}{2}$
$3\frac{1}{16} \times 3\frac{1}{16}$	H × H	5.000	$23\frac{1}{2}$	47	$22\frac{3}{4}$	2.523	2.523	$5\frac{3}{64}$	$22\frac{1}{2}$

45° Mini-Dual Valve

A Bore Size	GRAY-LOC Hub	MSP (psi)	B ℄ to Top	C Wheel to Wheel	F Over-all Length	R_1	R_2	℄	Number of Turns
$2\frac{1}{16} \times 2\frac{1}{16}$	F × F	3.000	$18\frac{7}{8}$	$27\frac{1}{4}$	17	1.773	1.773	$3\frac{35}{64}$	15
$2\frac{1}{16} \times 2\frac{1}{16}$	F × F	5.000	$18\frac{7}{8}$	$27\frac{1}{4}$	17	1.773	1.773	$3\frac{35}{64}$	15
$2\frac{1}{16} \times 2\frac{1}{16}$	XF × XF	15.000	$21\frac{1}{2}$	$34\frac{1}{2}$	22	1.773	1.773	$3\frac{35}{64}$	$17\frac{1}{2}$
$2\frac{9}{16} \times 2\frac{9}{16}$	G × G	3.000	$20\frac{5}{8}$	$31\frac{3}{4}$	$18\frac{1}{4}$	2.000	2.000	4	18
$2\frac{9}{16} \times 2\frac{9}{16}$	G × G	5.000	$20\frac{5}{8}$	$31\frac{3}{4}$	$18\frac{1}{4}$	2.000	2.000	4	18
$3\frac{1}{16} \times 3\frac{1}{16}$	H × H	3.000	$23\frac{1}{2}$	37	$22\frac{3}{4}$	2.523	2.523	$5\frac{3}{64}$	$22\frac{1}{2}$
$3\frac{1}{16} \times 3\frac{1}{16}$	H × H	5.000	$23\frac{1}{2}$	37	$22\frac{3}{4}$	2.523	2.523	$5\frac{3}{64}$	$22\frac{1}{2}$

All dimensions are in inches.
Source: Gray Tool Company

To install the maximum number of wells with good access and practicality, special clearance connections are used wherever possible. The Grayloc® clamp, which replaces flanged connections, offers greatly reduced size and weight, makes up faster and easier, and can be positioned at any angle without limitation by bolt hole alignment (Fig. 8–2). The space-saving potential of the clamp is a major consideration, as is its ability to make up without having to align flange bolt holes or have access from one side only.

Selecting a wellhead equipment manufacturer with the experience to carry out a multiwell through-the-leg project is an extremely important factor in projects of this type. This ensures the specialized equipment configurations and connections are already tried and proven, and that methods and procedures have benefited from actual oilfield installations.

While standardization of completion components is desirable for interchangeability and proven performance, ease in installation and possible later changes in well or equipment programs dictate that flexibility be designed into the system. This requirement may, in some instances, force the use of

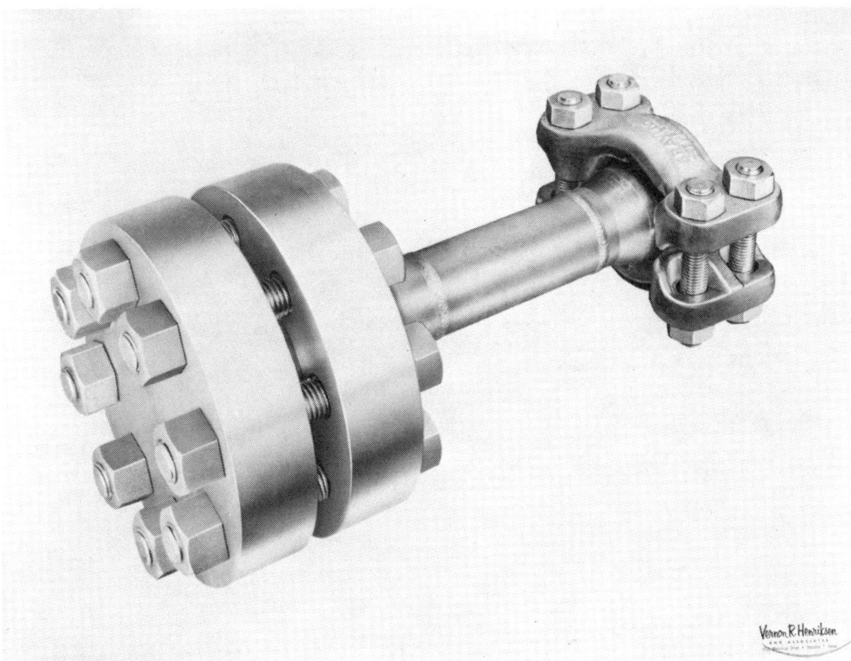

FIG. 8–2 A Grayloc clamp.

nonstandard hardware or procedures. Reliability of the well-control system should never be compromised, however, which is another reason for both selection of a wellhead manufacturer with through-the-leg experience and the thorough utilization of the systems design concept.

PROCEDURES

Generally speaking, completion procedures for through-the-leg programs are identical to those of any similar well in a normal platform deck completion. As mentioned, downhole equipment, casing, tubing, and controls are the same, and topside hardware generally varies only in configuration (see Chapter 7).

Through-the-leg wells usually have a more detailed and specific set of running and hookup procedures than conventional platform completions. The necessity for a proper equipment and piping installation sequence stems from the close working quarters in a concentrated well cluster, not from new or exotic hardware. Access to make up connections, test equipment, set packoffs, and similar operations must be maintained throughout the assembly. Piping and hookup tolerances and procedures will likewise be tighter than on a conventional platform completion and should be detailed as part of the systems design process.

9

Tension-Leg Completions

Today's demands for early production and rapid exploitation of offshore oil fields as a means of accelerating returns on the huge investments in exploration and development has brought a new urgency to the design and application of innovative offshore production systems. Meanwhile, accelerating costs have placed even greater emphasis on broader utilization of proven, standard equipment and techniques. The tension-leg platform (TLP) concept appears to offer a cost-effective approach for filling these needs.

The TLP offers a great deal of flexibility in completion methods. A unique wellhead tieback system, designed for use with the tension-leg platform (see Chapter 4) presents achievable capabilities for TLPs by incorporating conventional land-type equipment and methods, including surface trees. However, the tension-leg platform also readily interfaces with subsea production systems using standard floating drilling techniques (see Chapter 3).

TIEBACK SYSTEM

Each of the major components of the tieback system contributes uniquely to tie back each subsea well to the platform for control at deck level:

- *The alignment funnel remotely guides the tieback assembly onto the subsea wellhead housing, engaging a threaded connector that preloads the AX gasket to form a seal against internal well pressure. This connection is designed to withstand the maximum combinations of riser tension, internal pressure, and bending moments expected without overstress and fatigue.*
- *The tapered stress joint at the base of the riser also is utilized to minimize stress and increase fatigue life.*

- *Integral connections of the production riser are stronger in both tension and bending than the pipe body of the riser.*
- *Gimballed unitized tensioners keep the individual well risers in tension through platform excursions due to current and wave forces and thermal expansion and contraction.*

The TLP tieback system features individual risers to conduct production fluid from each well to the TLP production platform (Fig. 9–1). Designed for a diverless installation, the individual production riser systems are retrieved easily for inspection and maintenance.

For the most part, this production (or tieback) riser system consists of existing designs or adaptations of present technology. The alignment funnel and stress joint form the lower riser package with intermediate tieback joints extending from the stress joint to the tieback riser tensioner. Atop and supported by the tensioner are the surface wellhead and production tree. The surface wellhead attaches to the upper end of the 13⅜-in. diameter tieback riser tensioning joint and supports the 9⅝-in. diameter tieback riser.

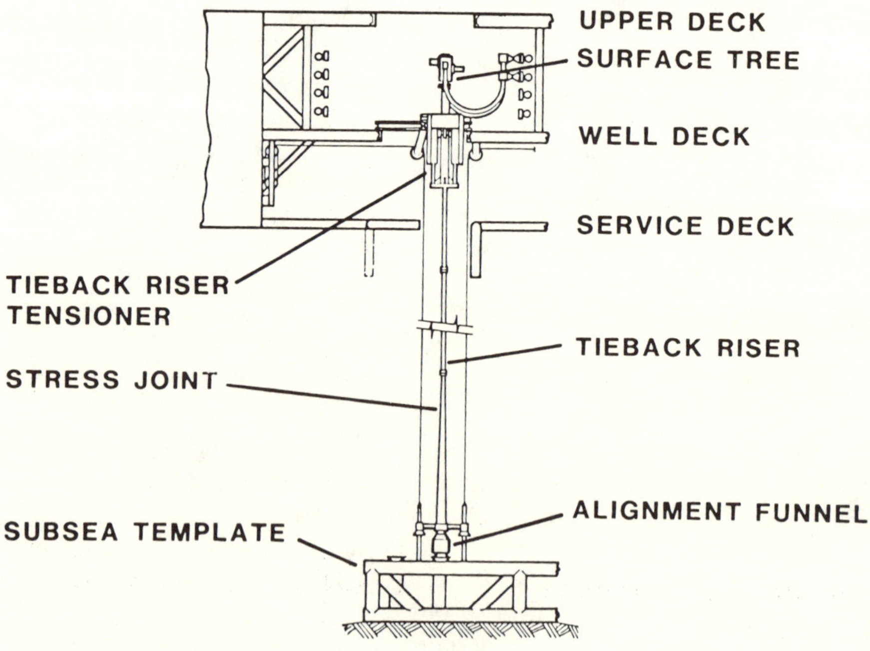

FIG. 9–1 Tieback system overview.

As with any direct tieback system, production tubing is suspended on the surface, providing direct annular access. The production casing bore is available to run downhole pumps and various other stimulation tools in case of heavy crude or other adverse reservoir characteristics.

$16^3/_4$-in. × $9^5/_8$-in. TIEBACK SYSTEM

This tieback system utilizes a single AX gasket as a primary seal. The tieback joint assembly, flanged to the bottom of the 9⅝-in. stress joint, includes tieback tool, torque sleeve, and alignment funnel (Fig. 9–2).

First, the funnel aligns the tieback joint as it enters the wellhead and guides the AX gasket into the casing hanger. Bending moments generated by the riser are transferred into the wellhead housing through the upper and lower alignment faces on the funnel. Reaction forces are reduced by the extra-long wellhead housing and guide funnel. The tieback joint assembly with guide frame is run down the guidelines to stab over the wellhead. Permissible angular misalignment is approximately 0.5°, slightly greater if the wellhead is not perfectly vertical. Setting weight on the tieback riser causes

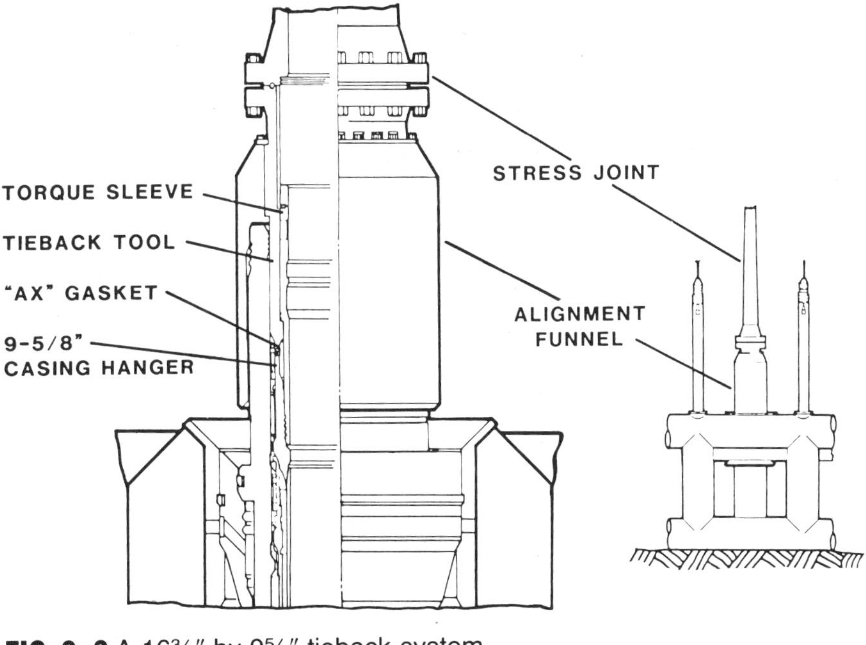

FIG. 9–2 A 16¾″ by 9⅝″ tieback system.

the tieback funnel to drop over and fully engage the wellhead. However, caution should be exercised to prevent dropping the full weight of the drill-string onto the hanger.

When the torque sleeve is lowered to the 9⅝-in. tieback casing hanger, right-hand rotation causes the threads of both to engage. Approximately 40,000 ft-lb torque are applied to give the desired preload and energize the AX gasket to effect a metal-to-metal seal.

PRODUCTION RISER TENSIONER

The production riser tensioning system is similar, both in concept and operation, to conventional hydraulic-pneumatic tensioners (Fig. 9–3). The system maintains the proper amount of tension on the riser assembly at all times, eliminating the possibility of buckling due to column length or breaking due to TLP motion.

The riser tensioner is rated at 445,000 lb. The unit's tension is developed through hydraulic cylinders operating in unison, each having a 6-ft stroke and a maximum hydraulic system pressure of 3,500 psi. The cylinders are grouped in diagonally opposite pairs, each pair manifolded to an air-oil accumulator. In turn, each accumulator is manifolded to an air backup vessel that supplies the added air volume to give the system a full-stroke load

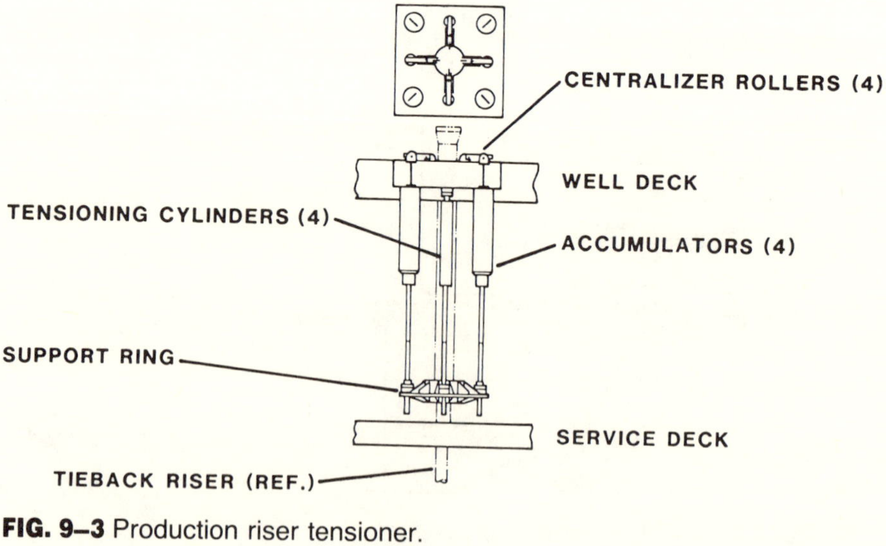

FIG. 9–3 Production riser tensioner.

variation of 25%. A control console with the valves and gauges necessary to vary or monitor tension loads is included in each tensioner module. Hard-piped air and makeup oil supply lines are required at TLP deck level.

The production riser tensioner allows passage of all production riser components (except the lower guide frame), easing installation. A special landing and tensioning joint at the upper end of the production riser facilitates landing in the tensioning ring. This special joint is actually a short piece of 13⅜-in. casing with landing preparations and gussets added for extra strength. This landing joint also allows the riser to be pulled more readily and helps support the production riser terminal head and surface production tree.

PRODUCTION WELL BAY

The primary support structure of the well bay carries the dead weight of each integrated tree/tensioner assembly and associated equipment (Fig. 9–4). The deck structure also supports the tension loads generated by each of the production riser tensioners and the sales riser tensioner. There are three main decks in the well bay: upper deck, well deck, and service deck.

Wide flange beams provide the basic support of the upper deck. These beams are arranged in a matrix that coincides vertically with the well slots at the template. The upper deck suspends the flow lines of the individual trees.

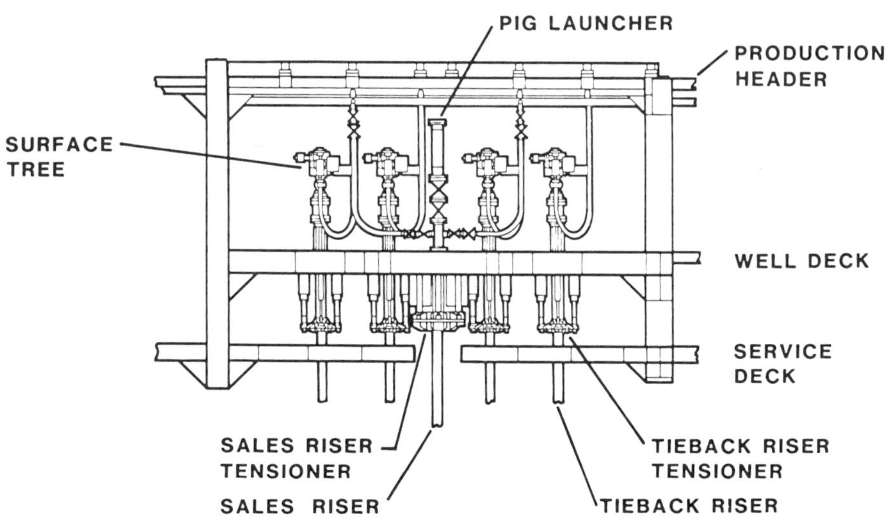

FIG. 9–4 Production well bay.

The middle or well deck also coincides vertically with the well slots and the other two decks. The opening for each well is designed so the modular production riser tensioners rest on inside supports with the top surface of the tensioners flush with the top of the deck. The sales riser tensioner is accommodated in the same way.

The service deck is used mainly for storage and for running and retrieving guide frames in production and drilling operations. In addition, this deck provides access for maintenance of the sales and production riser tensioners.

Other equipment installed in the well bay includes the surface wellhead/production trees and production/piping manifolding. The surface wellhead furnishes a landing for a surface tubing hanger. The tubing hanger, in turn, supports the weight of the tubing strings above the packer. Conventional surface production trees are totally compatible with this tieback system. A manual choke capability may be added if desired.

PRODUCTION RISER STRESS JOINT

The production riser stress joint is located just above the subsea wellhead. It has an unchanging bore throughout its 50-ft length but an outer diameter that varies linearly from approximately 12½-in. at the bottom to 9⅝-in. at the top (Fig. 9–5). This tapered profile provides uniform tensile and bending

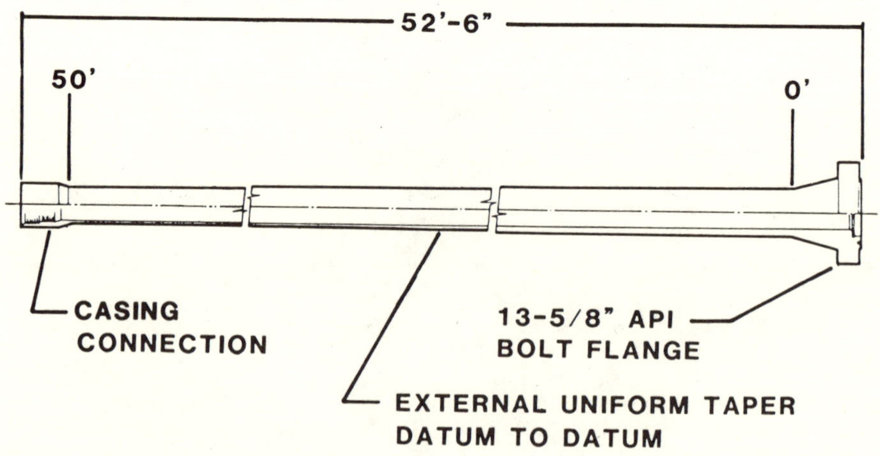

FIG. 9–5 Production riser stress joint.

stress over the joint's entire length. Consequently, angular rotation or bending of the lower section of the production riser is evenly distributed, and the curvature and associated stress levels can be predicted by computer analysis. This also solves the problems of bending the production and annulus tubing within the production riser.

The lower end of the stress joint is a 13⅝-in. diameter, 5,000-psi API flange for attachment to the tieback joint. The upper end has a high-strength threaded connection.

TLP SALES RISER

Sales risers offload the processed crude oil and natural gas from the tension-leg platform through subsea pipelines. As with the other risers, the sales riser system consists of the lower riser package, intermediate section, and top section. The basic differences between the sales riser and the high-pressure drilling riser are in the diameters and the provision for connection to a sales line pull-in head assembly at the template. The template structure has a pull-in receptacle that includes a landing ring for a 30-in. diameter anchor pile, an entry ramp for a pull-in head assembly, and locking profiles within the template structure to which a pull-in tool may be latched. Once the template has been landed and leveled and the production wells have been drilled, a 30-in. diameter anchor is installed through the landing ring of the pull-in receptacle. The pile serves as an anchor point for the subsea pipeline and as a reaction point for the tension load in the sales riser (Fig. 9–6).

The lower riser package is run at the bottom of the sales riser string and contains the equipment that locks the riser to the pipeline mandrel at the template. It also allows angular deflection of the riser with respect to the template due to surging of the TLP. The angular deflection allowance is accomplished using a flex joint flanged to the top of the high-pressure hydraulic connector on the pipeline mandrel. The whole assembly is run on the sales riser, using a guide frame to land and lock the assembly to the template.

The intermediate section of the sales riser system extends from the lower riser package at the sea floor to the well bay area of the TLP at the surface. Two integral hydraulic lines provide surface control of the hydraulic connector. The sales riser connectors are the same proloaded, metal-to-metal gasket-type seals used in the drilling riser. The joints are 50 ft long, using pup joints to space out as required.

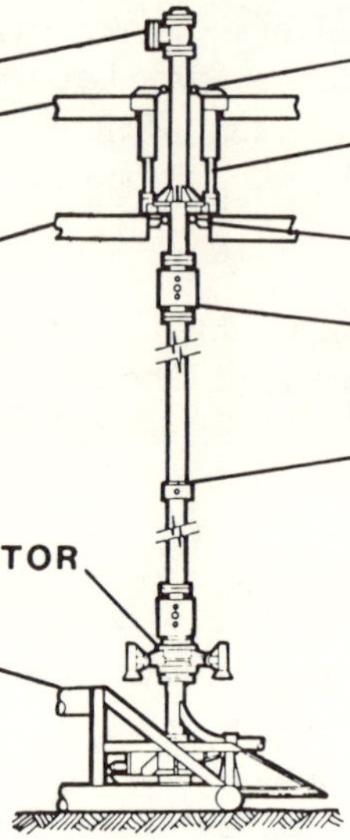

FIG. 9–6 TLP sales riser.

SALES RISER TENSIONER

The hydraulic/pneumatic sales riser tensioning unit is rated at 445,000 lb. The unit's tension is developed by six hydraulic cylinders, operating in tension, each with a 6-in. bore, 3-in. rod, 6-ft. stroke, and a maximum hydraulic system pressure of 3,500 psi. The cylinders are grouped in diagonally opposite pairs, each pair manifolded to an air/oil accumulator. Each accumulator is manifolded to an air backup vessel that supplies the added air volume to give the system a full-stroke load variation of 25%. The tensioner requires a single high-pressure line from the central tensioner monitoring-control panel.

The six cylinder rods are attached through gimbals to a common tensioning ring that reacts against a landing ring on the tensioning spool. The blind ends of the cylinders attach to the structural support frame from which the entire tensioner module is suspended.

The top section of the sales riser system includes the sales riser tensioner, tensioner spool, pig-launching equipment, and multiple-hose flex connection that attaches the terminal head of the sales riser to the collection manifold overhead. The terminal head has four horizontal outlets for attaching the flexible 4-in. ID hoses. Each hose may be isolated with shutoff valves that allow individual replacement without affecting overall production.

SUBSEA PRODUCTION SYSTEM

Another recent system designed by Vetco Offshore Inc. utilizes a subsea template with multiple production risers and a retrievable flow-line isolation valve manifold to produce 24 subsea satellite and template wells to a tension-leg platform in 1,500 ft of water in a hostile environment (Fig. 9–7). Many elements of the system's design represent new technology and, therefore, are preliminary by nature.

The riser system is composed of a template base and three central riser cores—for structural support—around which eight individual flow-line risers, one hydraulic hose bundle, one common annulus line, and a common service line are guided. Also included are a lower production riser package (LPRP) and a retrievable flow-line isolation valve package. Each producing well has its own flow path to the surface, which makes individual well monitoring quite simple.

The rectangular production riser base template carries pull-in structures that enable it to act as the gathering point for flow lines from all the satellite templates (Fig. 9–8). It also incorporates a preparation for receipt of the retrievable flow-line isolation valve package and a sales line connection. Principally, however, the riser template is the foundation for the three 12¾-in. diameter production risers. The riser assemblies are attached to the riser base by a hydraulic connector with the mandrel of the connector welded to the template itself. The template is designed to accommodate atmospheric diving systems (ADS).

Should a problem occur in a riser flow line or in the isolation (shutoff) valves, the retrievable flow-line isolation valve package allows 1) retrieval of the isolation valves without pulling the riser or 2) retrieval of the riser without pulling the valves, as must be done when valves are an integral part of the riser package. The compact through flow line (TFL) arrangement affords easy handling (Fig. 9–9).

A lower production riser package, located at the base of each riser, provides compensation for lateral movement of the TLP by incorporating a stress joint with wicker thread connections for flow lines. The wicker stabs by which flow lines are attached to the lower portion of the LPRP allows the flow lines to be installed without rotation—a real plus in deep water or rough weather.

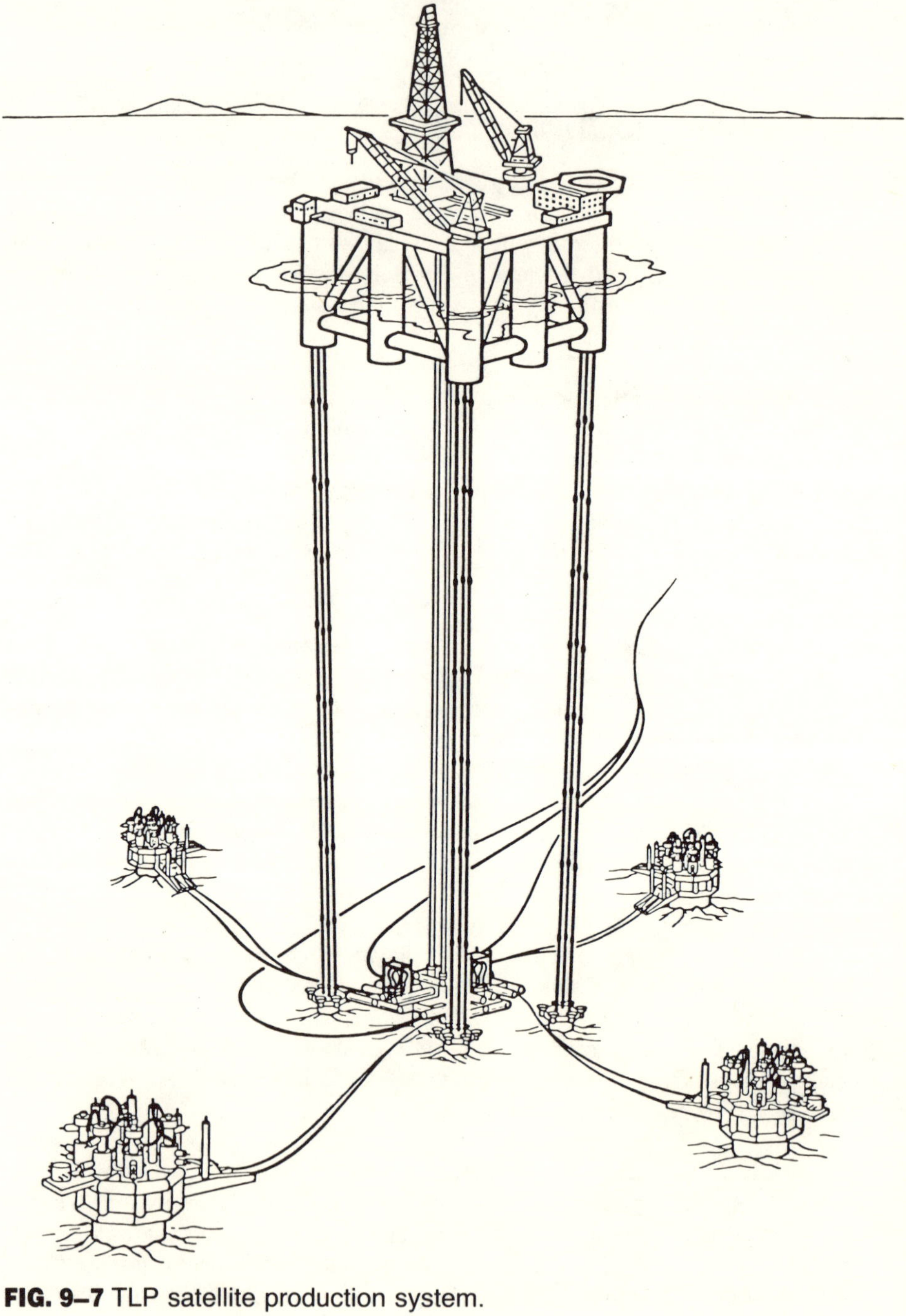

FIG. 9–7 TLP satellite production system.

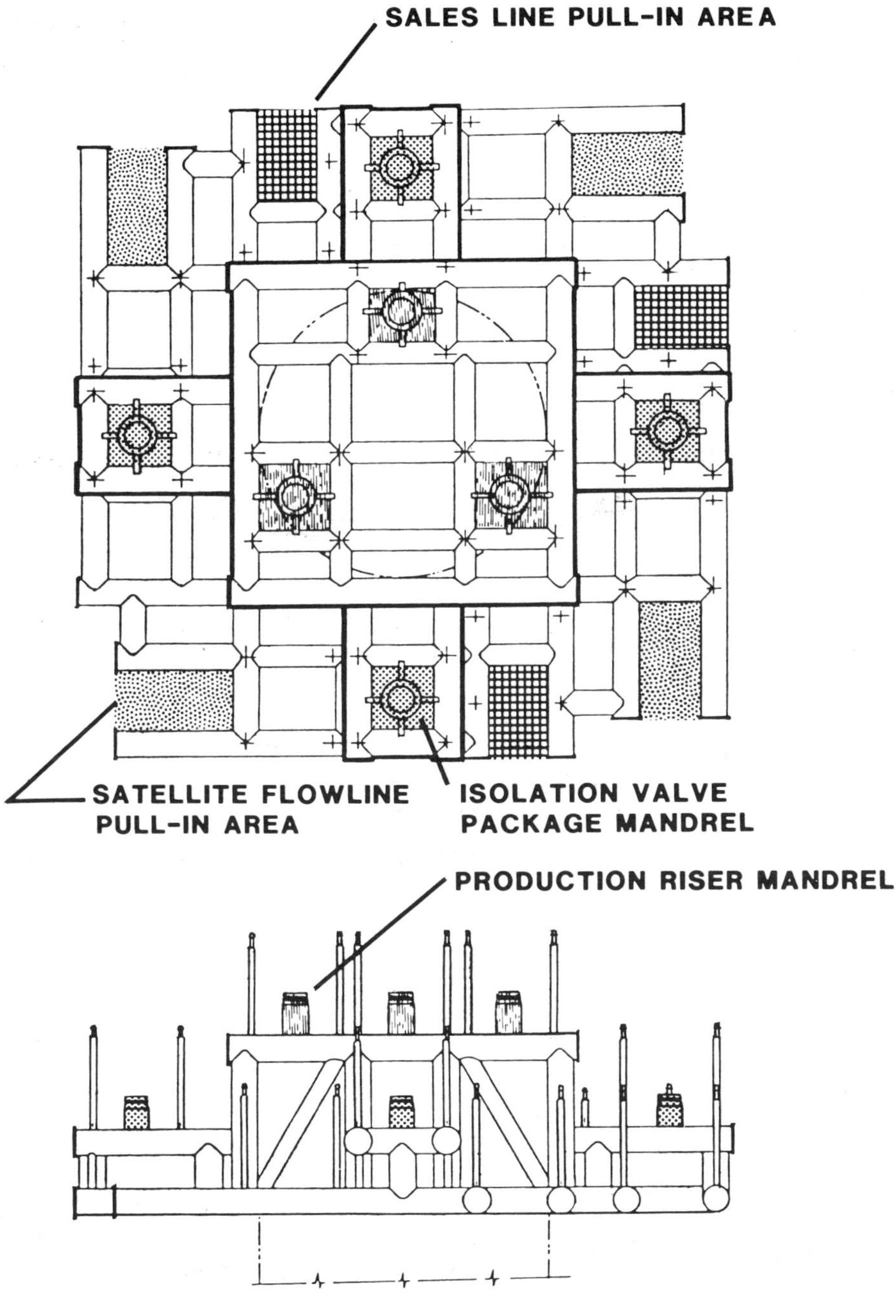

FIG. 9–8 Riser base template.

FIG. 9–9

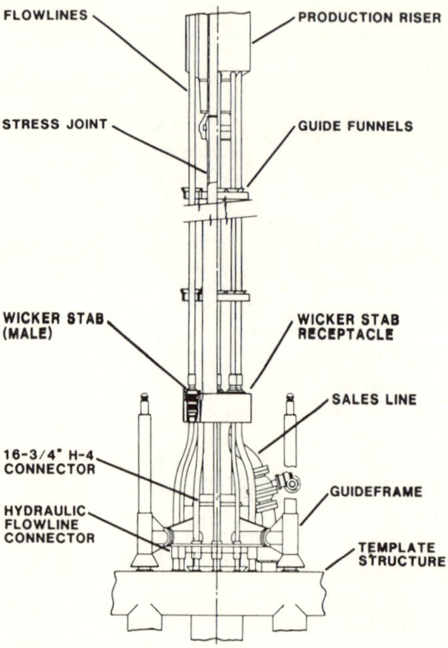

FIG. 9–10

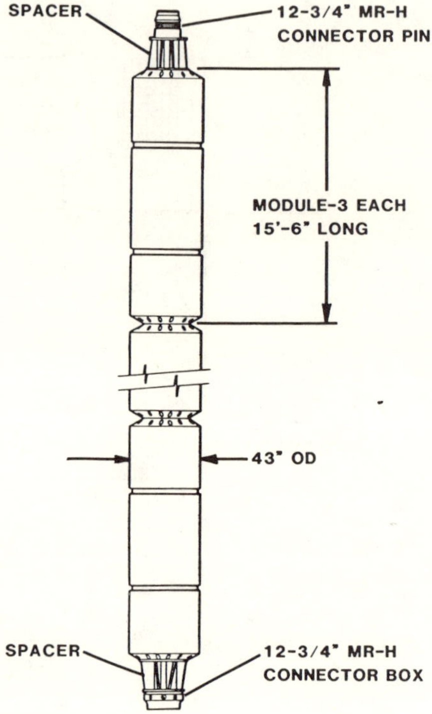

FIG. 9–9 Lower production riser package.
FIG. 9–10 Production riser joint with buoyancy modules.

Buoyancy modules around a production riser joint (Fig. 9–10) provide guidance for the flow lines and various other lines along the length of the riser, negating the need for guide cones and associated hardware. They also provide buoyancy to the riser package—4,000 lb per joint or approximately 100,000 lb buoyancy force per riser string. This reduces tensioning requirements by a like amount. In addition, the resultant lower drag coefficient afforded by the smooth surface structure reduces stress.

The same riser tensioning system used with the tieback is incorporated into the subsea production system to maintain the proper amount of tension on the riser assemblies at all times.

10

Multiwell Subsea Completions

During the 1970s and '80s the reliability and cost effectiveness of subsea well completions developed to a level that now makes production from so-called marginal fields economically feasible. A marginal field is defined as an offshore oil reserve that cannot economically support the installation of fixed drilling and/or production platforms.

The predominant multiwell subsea completion system utilizes a floating drilling vessel, which drills wells through a subsea template. Subsea trees are installed on the template, and wells are then produced through a production riser to a floating storage vessel or through flow lines to a remote tanker loading facility or platform. This system can also incorporate satellite trees connected to the template by subsea flow lines (Fig. 10–1).

This type of system offers a number of advantages, including early installation, since a template can be built and installed in less time than a fixed platform. The cost of the system also is not as sensitive to water depth as a fixed platform. Another advantage of multiwell subsea completions is flexibility, evidenced by the following features.

- *Accommodates a number of satellite wells that can be drilled while template drilling is underway or even before template installation.*
- *Most wells are located on a fixed sea-floor template.*
- *Can be installed in water depths of 1,000+ ft.*
- *Major components can be retrieved for reuse.*
- *Final location of the template can be determined after the reservoir has been defined by delineation drilling.*

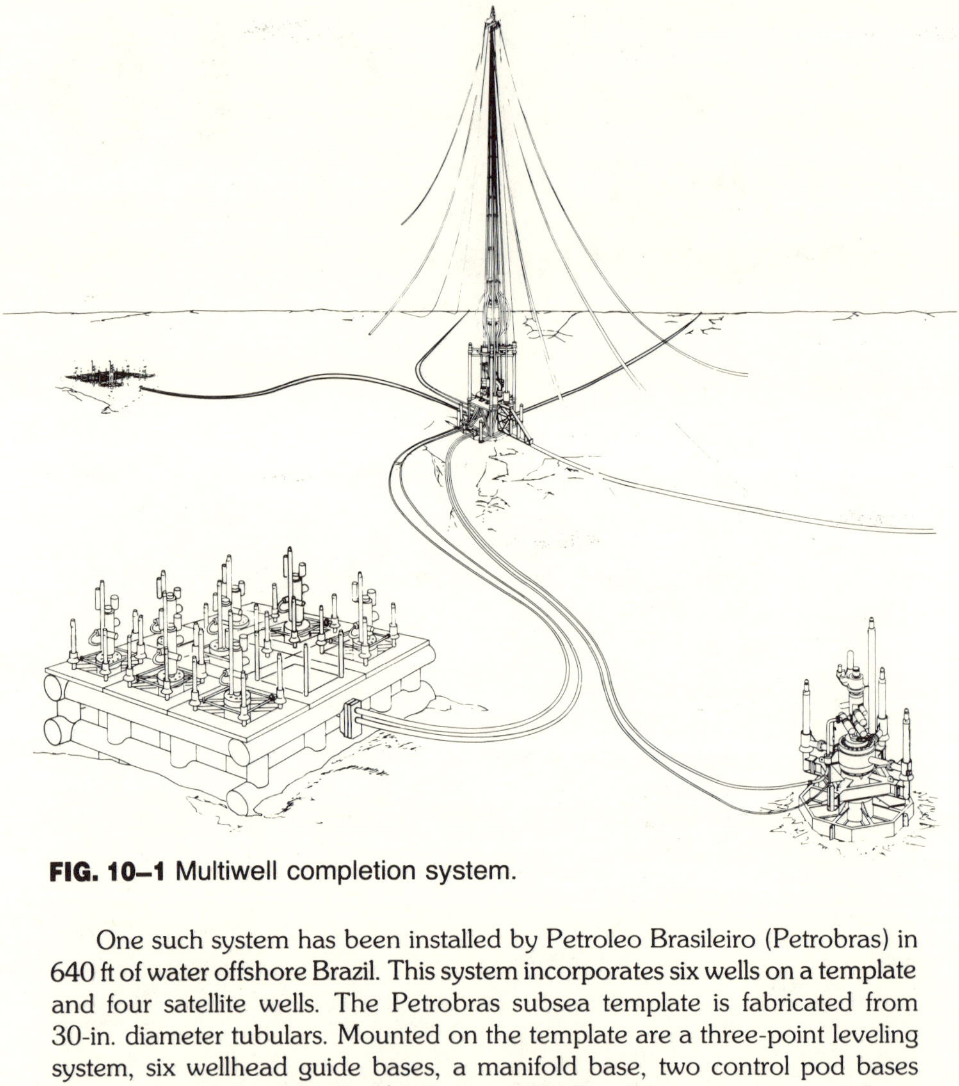

FIG. 10–1 Multiwell completion system.

One such system has been installed by Petroleo Brasileiro (Petrobras) in 640 ft of water offshore Brazil. This system incorporates six wells on a template and four satellite wells. The Petrobras subsea template is fabricated from 30-in. diameter tubulars. Mounted on the template are a three-point leveling system, six wellhead guide bases, a manifold base, two control pod bases (one electric and one hydraulic), four satellite flow-line connections, and two sales line connection receptacles. Interconnecting piping and control lines are also mounted on the template (Fig. 10–2).

The leveling system allows the template to be adjusted from an initial inclination of up to 5° to a final position of within one-half degree of horizontal. A level template is required to eliminate excessive wear on wellhead and drillstring components during drilling, to minimize bending moments and forces transferred to the template during drilling, and to facilitate engagement of subsea stabs. The leveling system uses gimballed slip devices to support the template on three 20-inch OD piles previously cemented into the sea bed (Fig. 10–3).

The wellhead guide bases are modified from standard permanent guide structures to permit passage of a 36-in. diameter bit. Each guide base is equipped with remote post tops on each of the four posts and vertical stabs for connecting the flow line, annulus line, hydraulic lines, and electrical cable from the template to the tree.

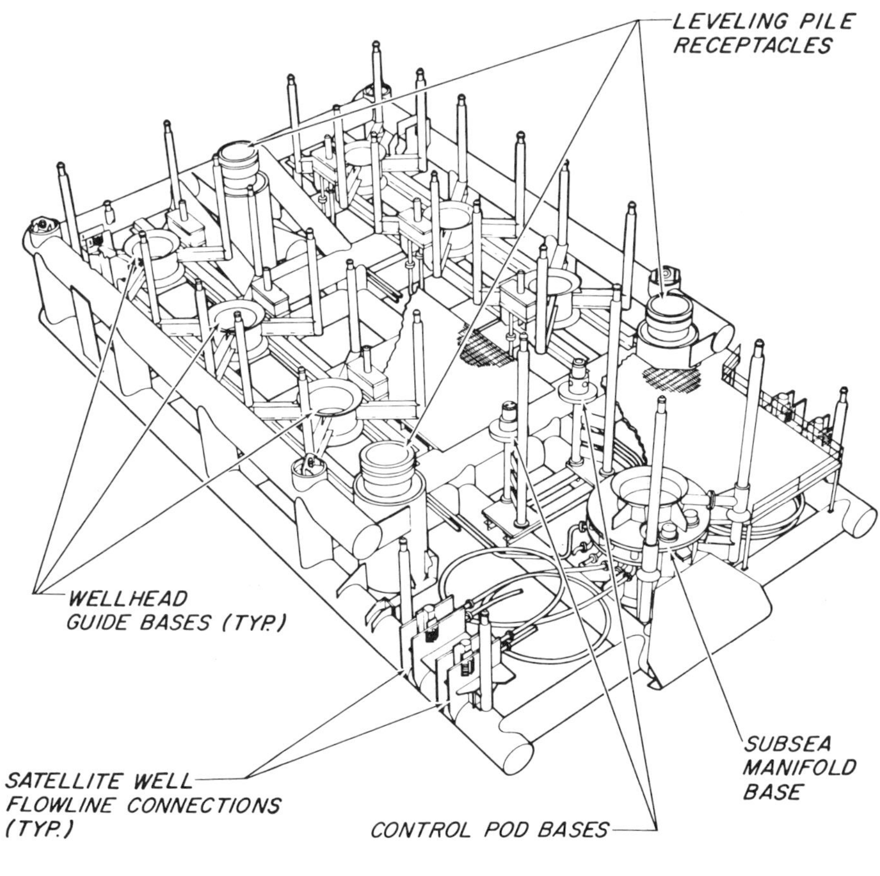

FIG. 10–2 Petrobras Enchova template.

The subsea manifold base provides a vertical stab for the flow lines and annulus access lines from each of the ten wells. It also has a receptable for landing and cementing a 30-in. diameter pile to which the manifold is later attached with a hydraulic connector. Two receptacles are incorporated for receiving the vertical stab on the end of the 8-in. sales lines that connects to the tanker loading terminals. The control pod bases, located adjacent to the manifold base, are attachment points for hydraulic and electrical pods for controlling the ten subsea trees.

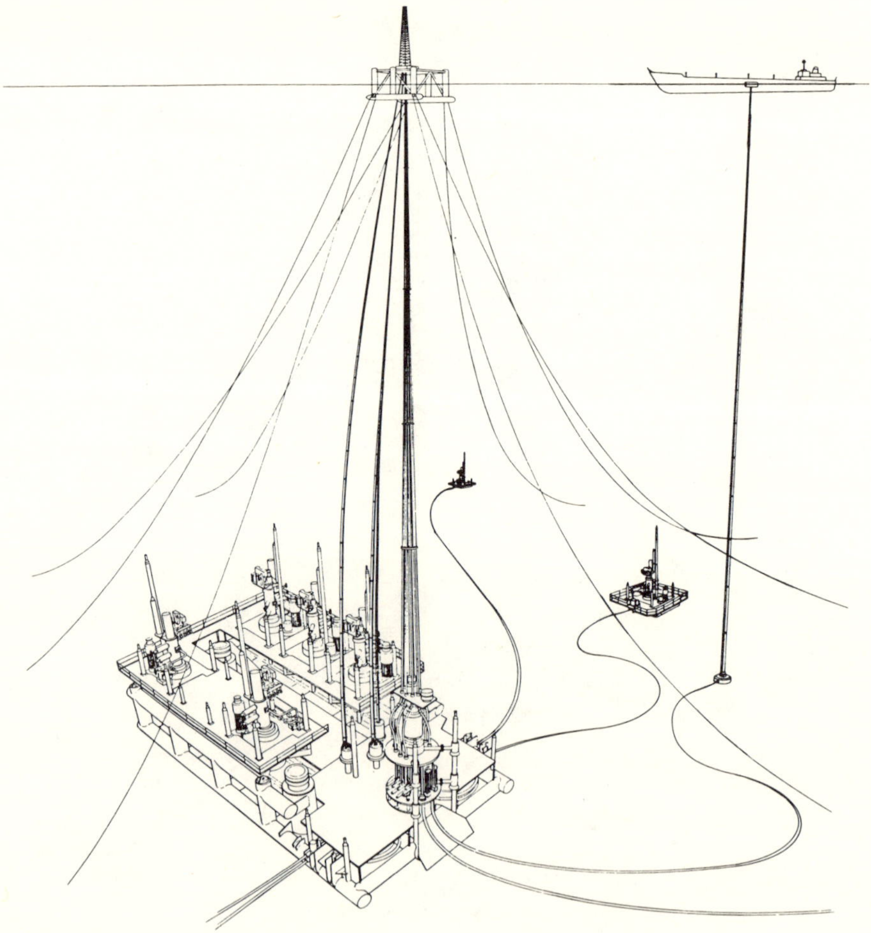

FIG. 10–3 Drilling from a floater through the template.

The four satellite flow-line connections join the flow line, annulus line, hydraulic supply line, and electric control cable from the satellite wells to the subsea template. These diverless flow-line connections have retrievable metal-to-metal seals.

The template-mounted piping connects the vertical stabs at the wellhead guide bases with the manifold base and connects the flow-line receptacles from the four satellite wells to the manifold base. The template-mounted control lines connect the guide bases and the flow-line connections with the control pod bases.

During drilling, protectors are used on the satellite flow-line connections, the control pod bases, the vertical stabs of the wellhead guide bases, and the manifold base.[1]

DRILLING

Once the template has been landed on the ocean floor and is leveled, drilling operations for each well slot are carried out from a floating drilling vessel in a conventional manner. Guidelines are established to fixed guide-posts on the template, which serves as the permanent guide structure. The drilling operation and tubing hanger installation are carried out in the same manner as discussed for a satellite tree (see chapter 11). Each wellhead system is landed and supported in an individual receptacle on the template.

TEMPLATE TREES

The template trees are nearly identical to the satellite trees with the exception of the flow-line connection. A vertical stab connection is made to the flow line and control line bundle as the tree is landed. The flow-line connection is latched with a hydraulically actuated connector. After the trees are installed, the next step is installing the production riser system.

PRODUCTION RISER SYSTEM

A semisubmersible drilling vessel (which has been converted for production) is anchored over the template for use during installing and attaching the

[1]L.E. Reimert and P.J. Gray, "A Ten-Well Subsea Production System for the Enchova Field," OTC Paper No. 3448, 1979.

production riser system (Fig. 10–4). Free-hanging flexible risers are attached to the semisubmersible's lower hull through quick disconnects. These attachments are made on the outer perimeter of the vessel for easy access and to leave the moonpool clear for running and retrieving subsea trees, tree components, or even a BOP stack and riser as required for well workover.

The system is comprised of five flexible risers:

- *8-in. diameter combined production line from template to production vessel*
- *8-in. diameter export riser from production vessel to remote tanker loading buoy (SBM)*
- *4-in. diameter annulus access and flow test line from template to production vessel*
- *4-in. diameter water injection line for one template well*
- *4-in. diameter water injection line for second template well*

The 8-in. diameter combined production line riser is pulled in and anchored directly to the subsea manifold base on the template. Once installed, the 8-in. diameter subsea termination presents an 8-in. diameter vertical stab for engagement by the manifold when it is lowered and connected in place.

The water injection line riser for one of the injection wells and the annulus access flow test riser will be connected to the template through a satellite well connection point.[2]

SUBSEA MANIFOLD

The subsea manifold serves two purposes: it gathers the production from seven wells and it connects the annulus lines from nine wells into a common header, which also serves as a flow test network (Fig. 10–5). Each production flow line is equipped with a 3-in. choke isolated by 3-in. gate valves for service access. Crossover networks on the satellite wells permit production through an annulus flow line if the normal production line is damaged. This network also permits flow testing the satellite wells through the normal production flow line to the manifold, where the fluid is diverted into the flow test network.

The template wells may be flow tested individually through the annulus flow line to the annulus header and manifold network. The subsea manifold

[2]*Petroleum Engineering*, March 1981, D. Gray

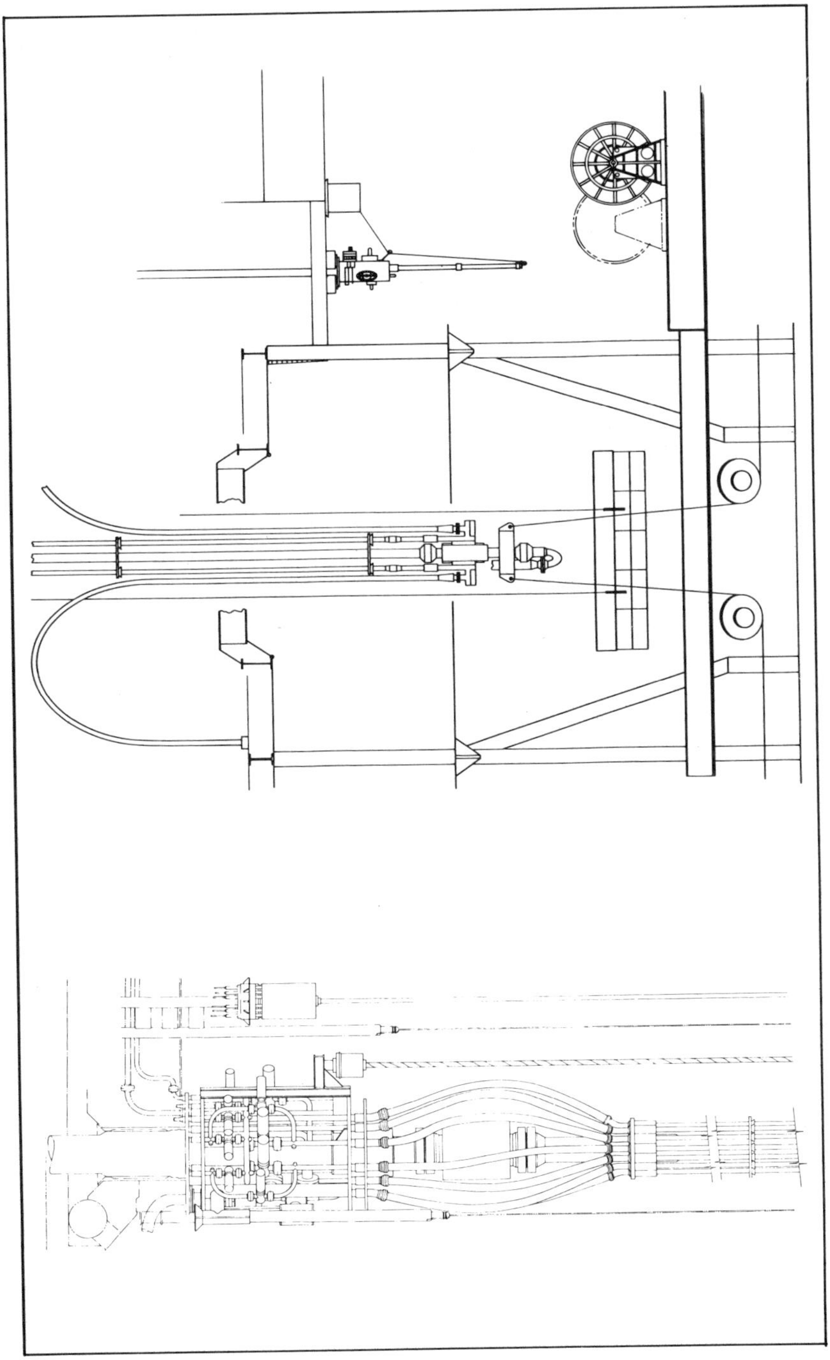

FIG. 10–4 Production riser system.

FIG. 10–5 Subsea manifold.

can be run at any time after the 8-in. diameter combined production line and water injection line risers have been connected to the template. (These lines present vertical stab connections to engage the manifold.)[3]

[3]*Petroleum Engineering*, March 1981.

11

Subsea Satellite Completion Systems

Single-well completions, commonly referred to as satellite completions, are used effectively in offshore field development in a number of ways. Exploratory and developmental wells can be completed with a subsea tree and produced by means of flow lines to a subsea manifold or to a centrally located platform. Satellite wells can be used on the outer fringes of shallow fields that are not accessible by means of deviated drilling from a platform or subsea template. Satellite wells are also used for injection purposes to enhance production. In fact, a system recently proposed by British Petroleum, single-well offshore production system (SWOPS), would allow intermittent production to a dynamically positioned tanker through a production riser lowered from the tanker and connected to the top of a sea floor satellite tree.

In the face of heavy up-front field development costs, the earlier a field can begin earning revenue, the sooner it can offset the total cost of this investment. Petrobras employed an early production technique producing single satellite development wells in the Enchova Field off Brazil through flexible flow lines to an anchored semisubmersible production vessel. This was an interim measure, allowing limited production until a subsea multiwell template and a permanent platform could be installed.

Most satellite trees installed to date are wet and, by definition, are exposed to the sea around them. There are some dry systems in existence that enclose the wellhead and production tree in a one-atmosphere inert gas-filled chamber.

Satellite completion systems can range greatly in degree of complexity, the simplest being a single-bore stacked valve tree with diver-assist flow-line connection and direct hydraulic control to a minimum number of valves. Complexity is increased with such features as multiple bores, through-flow line (TFL) workover capability, diverless installation, multiflange flow-line connection, electrohydraulic controls and monitoring systems, and guidelineless reentry.

WET SATELLITE COMPLETIONS

Through 1980, the number of different designs of wet satellite trees installed throughout the world was more than one-half the total number of wet trees. As the wet satellite tree concept has developed, designs have become more standardized in relation to the number of wet trees now in service. Regardless of detailed design, each subsea satellite completion system is comprised of the following major subsystems:

- *tubing hanger system*
- *completion riser*
- *production tree assembly*
- *flow-line connection system*
- *control system*
- *installation and workover system*

The example system described here is one of the satellite trees installed by Petrobras in the Enchova Field off Brazil. This is a 4-in. × 2-in.-diameter, 5,000-psi WP, non-TFL diverless tree with remote reentry capability (Fig. 11–1). The tree is controlled by an electrohydraulic system. Although it is designed to be installed and operated without divers, its design allows access by an atmospheric diving system (ADS-JIM suit) for observation, maintenance, and repair if required. Gate valve operator/gate seat assembly removal and reinstallation by JIM was demonstrated at an underwater test facility prior to offshore installation. Following is a basic description of each of the subsystems and its function.

Tubing Hanger System

Most tubing hanger systems are adaptable to a variety of tubing programs and production schemes. These tubing hangers lock and seal inside the uppermost casing hanger body in an existing wellhead installation. The tubing hanger and attached tubing strings normally are run as a complete assembly. Tubing hangers generally are available as either mechanical-set or hydraulic-set units. A significant advantage of the hydraulic-set tubing hanger is that, when run on the completion riser, it can be run, landed, sealed, and all downhole operations performed in a single trip (Fig. 11–2). This is possible

FIG. 11–1 Petrobras Enchova tree.

since an unrestricted throughbore connection is maintained and control ports (within the hanger body) for downhole functions are tied back to the surface through the running tool as part of the completion riser. Also incorporated in the hanger body are threaded inserts or internal profiles for landing and setting tubing plugs.

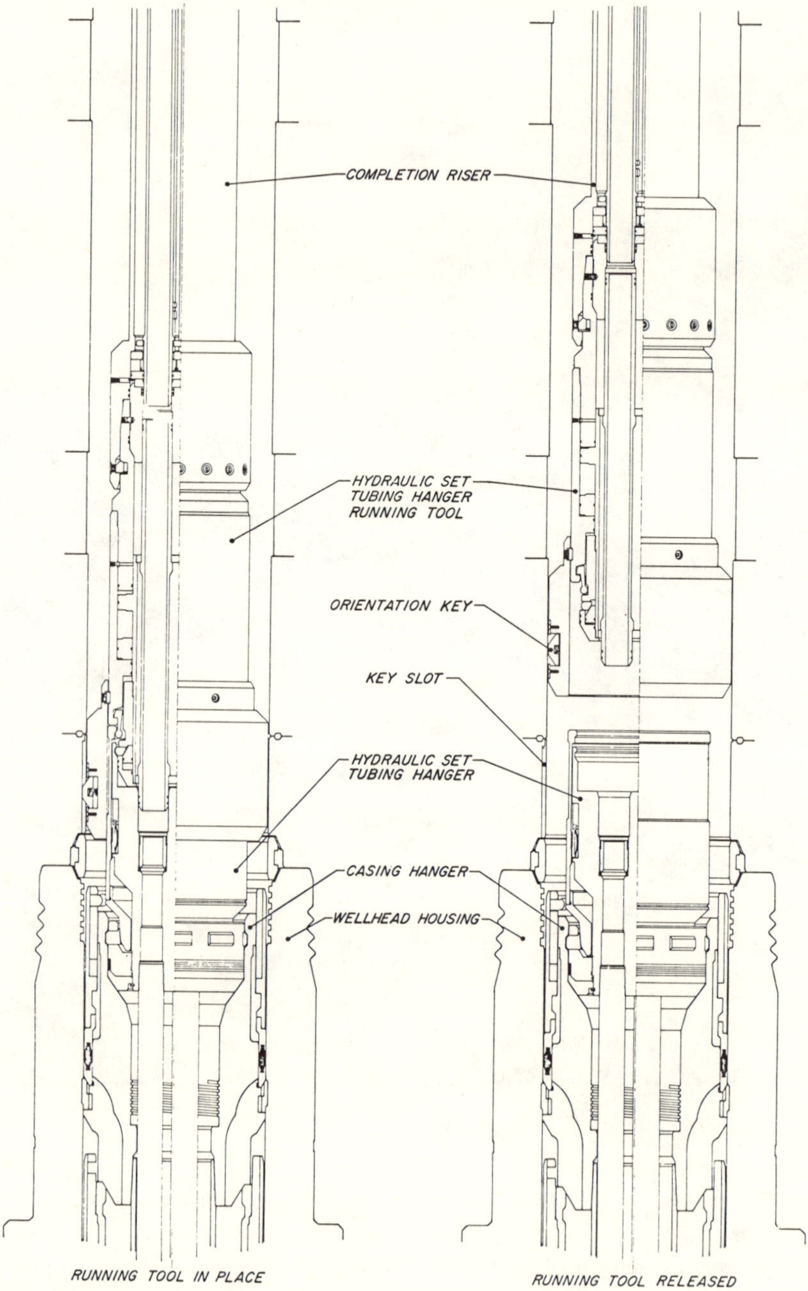

FIG. 11–2 Hydraulic set tubing hanger. *(Courtesy Vetco)*

The tubing hanger selected may provide for suspending a single string or multiple strings of tubing. The use of two or more strings enables production from separate zones and also permits downhole maintenance using pump-down TFL tools. The tubing hanger illustrated is equipped with a 4-in. production bore and a 2-in. bore to provide access to the annulus between the tubing strings and the casing. This 2-in. bore can be used to monitor pressure in the annulus or for gas lift.

To provide a safety shutoff in case of damage to the tree or flow lines, fail-safe close (FSC) valves called storm chokes or subsurface safety valves are included in the tubing strings. These valves, usually ball valves maintained in the open position by hydraulic pressure, are run with the tubing or on a wire line. When hydraulic pressure is removed, either intentionally or accidentally, the valves move to the closed position to prevent any further flow from the well.

The hydraulic-set tubing hanger is run through the drilling riser and BOP stack on a running tool and completion riser. The tool functions, operated through a series of hydraulic control lines installed internally in the completion riser, are connected to surface controls through the terminal head assembly. The tubing hanger is landed in the casing hanger, rotationally oriented, and locked in place by applying hydraulic pressure. Hydraulic pressure, ported through the running tool, is also used to release the running tool from the tubing hanger. This same tool retrieves the tubing hanger should it become necessary to pull the tubing strings.

Completion Riser

Because of the relative complexity of the multiple-bore tubing hanger with downhole safety valve ports and the requirement for several individual hydraulic control functions to the hydraulic tubing hanger running tool, a simpler operation is achieved by running the tubing hanger on the completion riser. The completion riser, illustrated in Fig. 11–3, encloses all tubing bore extensions, subsurface safety valve (SSSV) control-line extensions, and hydraulic controls in an outer casing that provides a known profile for calculating riser stresses and top tension requirements.

The various tubing bore extensions, (SSSV) lines, and hydraulic lines are spaced precisely to match the spacing on the tubing hanger running tool and tubing hanger. The pin-and-box end connections provide a means of fast makeup on the rig floor by activating the actuating screws carried on the pin section of each riser joint. The completion riser can pass through a 13⅜-in.

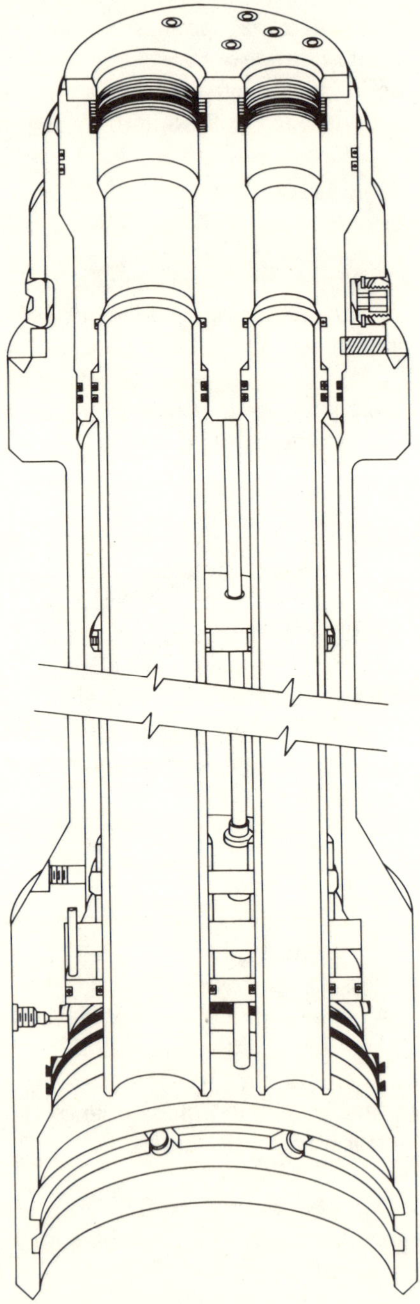

FIG. 11–3 Completion riser.

ID BOP stack and drilling riser system during installation of the tubing hanger. The same completion riser can also be used for production tree installation and workover.

Once the tubing hanger is landed and locked into place, downhole preparations are completed, and the system is successfully pressure-tested, blanking plugs are run on wire-line tools and set in the locking profiles of the tubing hanger. This is done so the well will be shut in effectively. Once the well is shut in, the tubing hanger running tool and completion riser can be retrieved and the BOP stack removed. A protective cover may be placed on the wellhead if the tree is not installed immediately.

At this point in the completion sequence of the well, the operator is given the choice of running the tree or the flow lines first. In this example, the tree will be run first.

The permanent guide structure for this system has the flow-line alignment structure attached as a permanent member. In those instances when it is decided to produce the well after the wellhead has been installed and preparations for flow line pull-in were not incorporated into the permanent guide structure, an alignment structure can be installed after wellhead installation.

Production Tree Assembly

The production tree is made of the following basic components necessary for controlling production fluids between the tubing hanger and the flow-line connection (Fig. 11–4):

- *tree guide frame*
- *wellhead connector*
- *valve block(s)*
- *wye spools*
- *swab valve block*
- *flow-line loops and crossovers*
- *tree manifold*
- *tree cap assembly*

Tree Guide Frame. The tree guide frame ensures positive guidance of the tree during installation. When the four guide funnels on the guide frame are stabbed over the guideposts of the permanent guide structure (PGS), proper rotational orientation is achieved (Fig. 11–5).

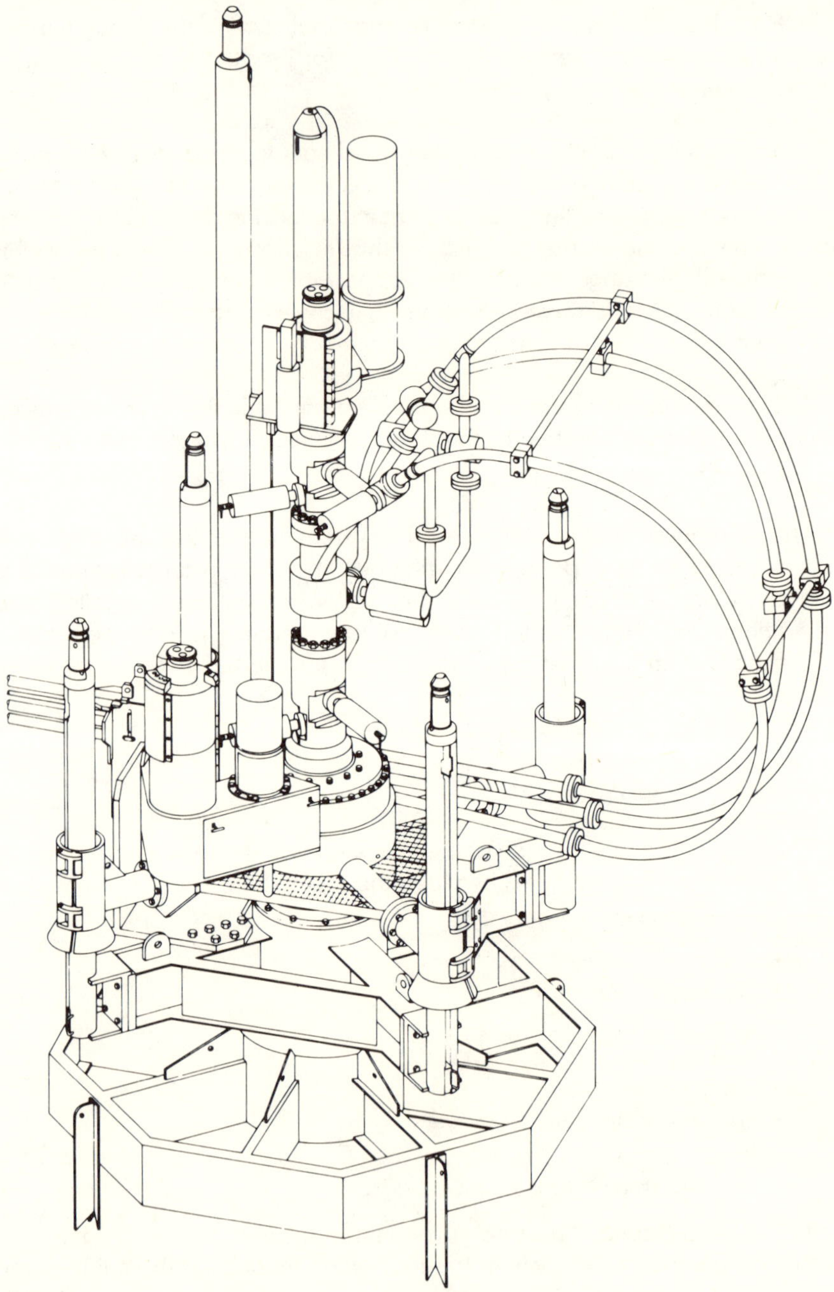

FIG. 11–4 TFL production tree. *(Courtesy Vetco)*

The heavy center ring of the four-funnel guide frame is bolted around the wellhead connector of the tree assembly to make the frame a permanent part of the tree. The funnels are flanged to the guide frame support arms and carry provision for attaching sacrificial anodes for cathodic corrosion protection. The tree guide frame is designed for diverless flow-line connection, but a walkway has been included to allow ADS servicing. Brackets on the frame provide structural support for the tree flow loops and production flange.

Wellhead Connector. The standard Vetco H-4 connector is a hydraulically operated unit that locks and seals the tree directly to the wellhead housing (Fig. 11–6). The connector is remotely operable, incorporating design features that allow it to combine its lock-and-seal integrity, essential during the life of the well, with a release-and-retrieve capability. When the connector is locked, the dog segments are securely maintained in their locked position against the wellhead housing profile by a cam ring positioned by hydraulic pistons. The surface between the cam ring and dog segments is a self-locking taper that, when the cam ring is moved downward to the lock position, forces the dog segments radially inward and locks them in place without the need to maintain hydraulic pressure on the cylinders. An indicator rod, attached to the cam ring, allows visual verification of the locking mechanism in lock or release position.

The connector top adapter has a studded-top flange for attaching the master valve block. The adapter has a stainless steel-lined preparation for the seal ring gasket. The interface between the connector and the wellhead is a metal AX or VX seal ring, rated at the working pressure of the wellhead and preloaded to ensure a reliable, pressure-tight seal.

Valve Block(s). For non-TFL trees, it is common to include the master valves (for primary control of the well) and the swab valves in a composite valve block, allowing a more compact tree configuration. This valve block mounts directly to the top of the wellhead connector and allows access to the matching tubing bores after the tree is installed. The valve block carries extension subs that remotely stab and seal in the tubing hanger. An orienting bushing bolted to the bottom of the valve block ensures accurate alignment with the tubing hanger (Fig. 11–7).

The tree illustrated has dual production master valves and a single annulus master valve. These valves, as well as the swab valves, are hydraulically operated FSC gate valves (Fig. 11–8). The number and type of master valves (manual or FSC) varies with the application and user requirements. In the

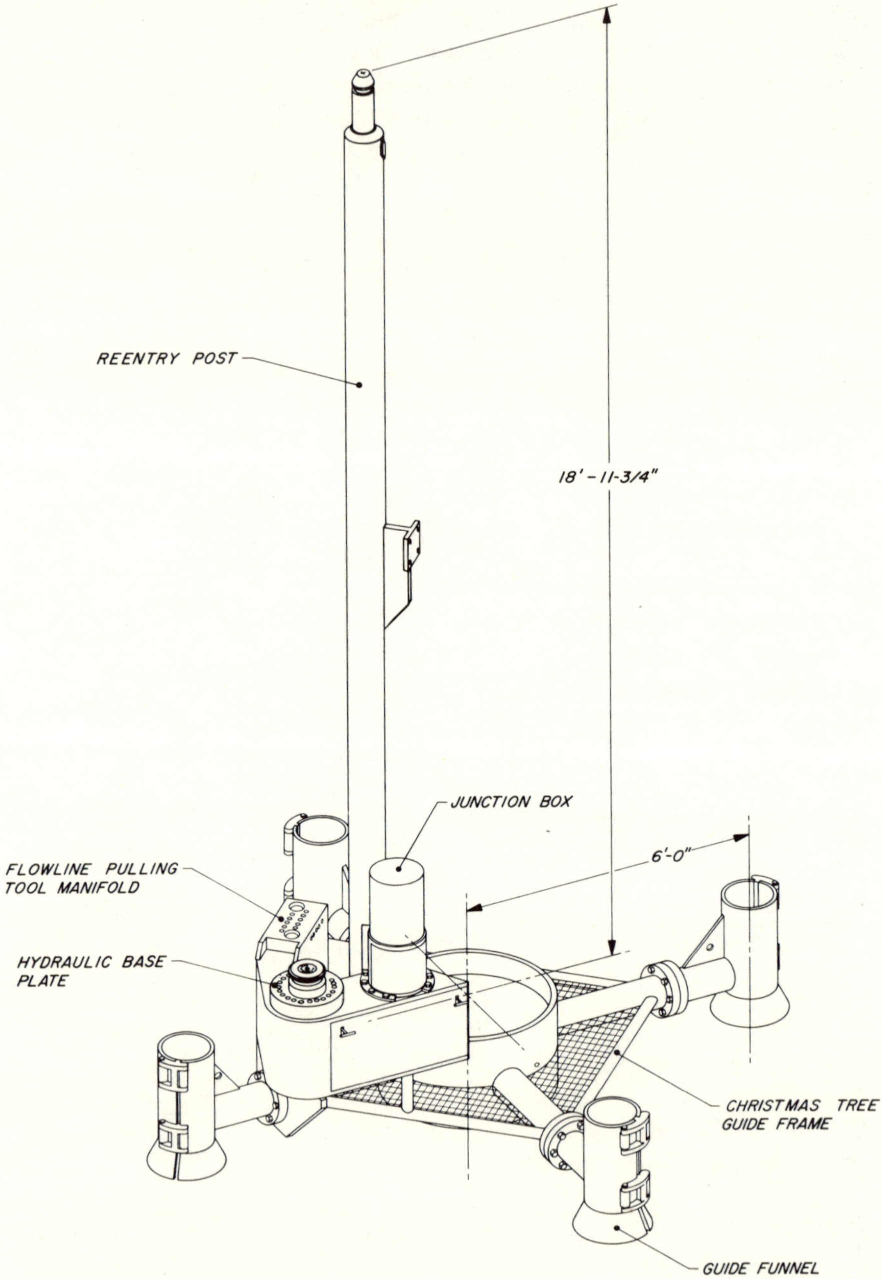

FIG. 11–5 Tree guide frame. *(Courtesy Vetco)*

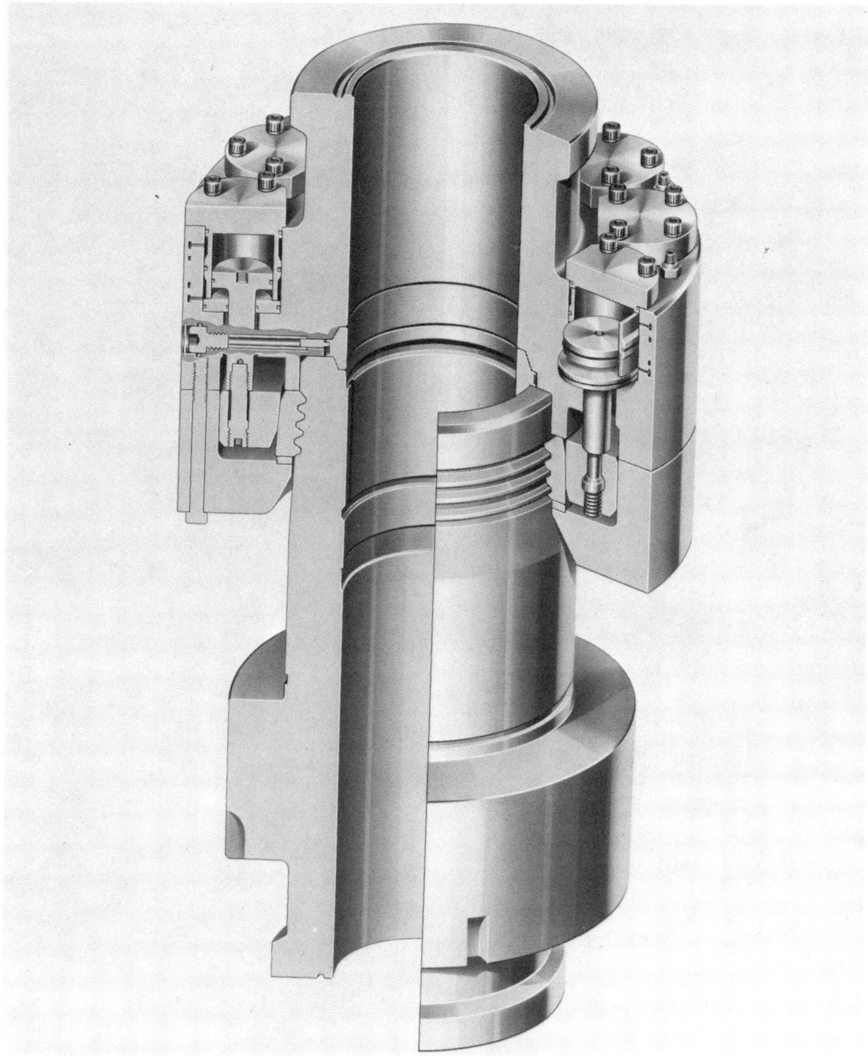

FIG. 11–6 Wellhead connector.

composite valve block, studded side outlets between the master and swab valves provide for mounting of the production and annulus wing valves, crossover valve(s), and piping spools.

Some users require that a weak point be designed into the tree to ensure the master valve block will remain intact should the top of the tree or the

FIG. 11–7 Composite valve block. *(Courtesy Vetco)*

flow loops be subjected to a high external force. This generally is accomplished by having a separate master valve block and a side outlet/swab valve block, with the weak point between the two.

Wye Spool. For TFL trees which are generally dual bore with an extra annulus access port, an additional component is required in the main body

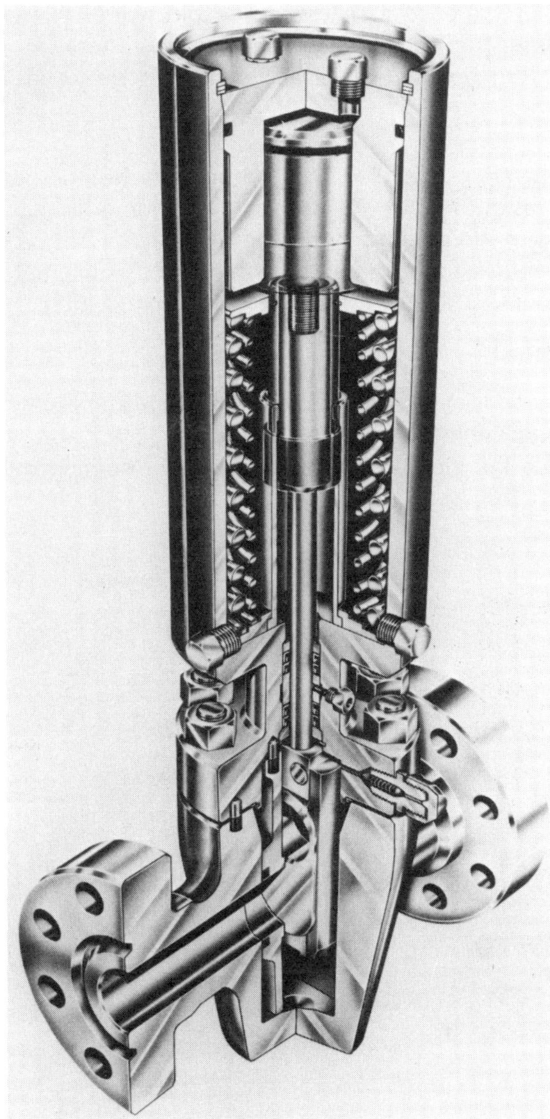

FIG. 11–8 Hydraulic FSC gate valve.

of the tree. This is a TFL diverter, more commonly referred to as a wye spool. The wye spool initiates the angle for the (minimum) 5-ft radius in both the service and production lines to accommodate TFL tools and service operations. It is mounted on the master valve block and allows both vertical and through-flow-line (TFL) access (for pumpdown tools) to the production and service tubing bores. Vertical access only is provided to the annulus bore.

Plug-type diverters, installed using wire-line techniques, are inserted into the vertical bores at the wye junction to provide smooth passage of tools to and from the downhole tubing bores through the flowline loops. Optionally, the wye spool may be fabricated with other types of integral diverter mechanisms that would be actuated either hydraulically or by passing the tool in conjunction with a detent mechanism.

Swab Valve Block. This provides vertical access to the well (Fig. 11–9). Operated only through the tree running tool, the FSC swab valves normally are closed when the well is producing. They are opened only during well-

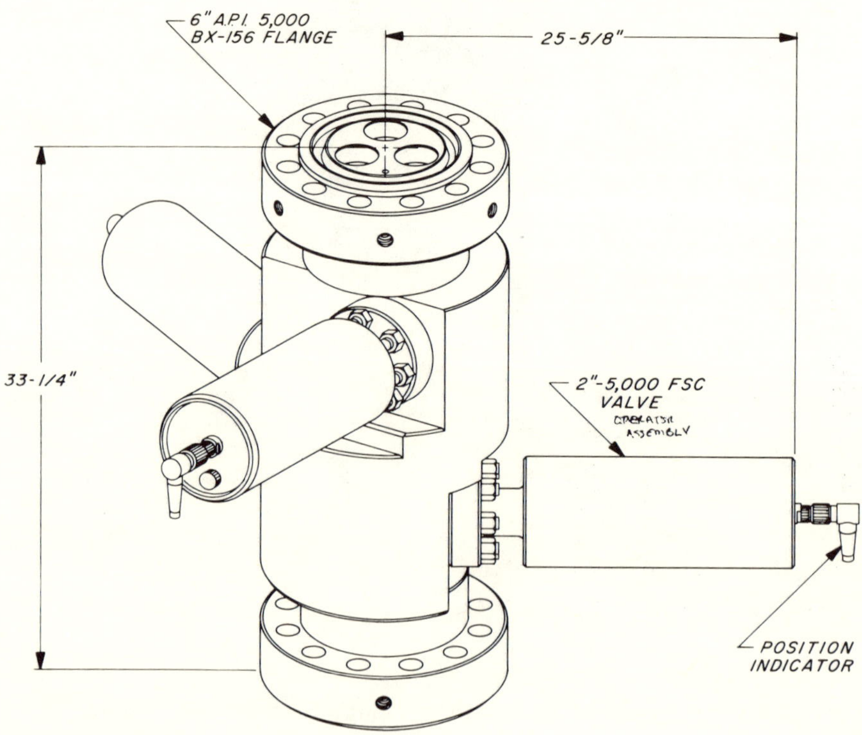

FIG. 11–9 Swab valve block.

servicing operations when vertical access to the tubing bores is required. The valves are identical to those used in the master valve block assembly.

Flow-Line Loops and Crossovers. The flow-line loops interface between the valve block and tree flange, providing flexibility for the necessary linear movement of the tree flange for engagement or disengagement of the flow-line connection.

Located in the flow-line loops are the wing valves. These provide individual fail-safe control for each flow line. When servicing the well vertically, the wing valves are closed and the swab and master valves are opened. A crossover line fitted with a fail-safe open (FSO) gate valve interconnects the flow line and service line. Additionally, with the wing valves closed, the cross-over valve can be opened to allow pigging of the flow lines.

Tree Manifold. The tree manifold is the uppermost fixed member of the subsea completion tree, mating directly to the top flange of the (composite) valve block (Fig. 11–10). The manifold includes the following features:

1. Throughbores installed using wire-line methods with internal preparations for receiving plugs
2. An external locking profile that accepts the tree cap or the tree running tool
3. A mounting plate for supporting the reentry post
4. Threaded ports for attaching the hydraulic lines controlling the tree functions
5. Receptacles for the hydraulic stab subs for either the tree cap or the tree running tool
6. Check valves, incorporated in the manifold, to isolate the hydraulic system from contamination when the stab subs are removed

The manifold is the terminal point for all of the hydraulic control lines. When the subsea installation is being worked on from the surface, the through-bores provide communication with the well bore through the completion riser and tree running tool.

The tree manifold, in conjunction with the tree running tool or the tree cap assembly, directs the hydraulic control functions of the tree as follows. When running or landing the tree, the tree manifold ports the hydraulic operating pressure directly from the tree running tool to the tree functions.

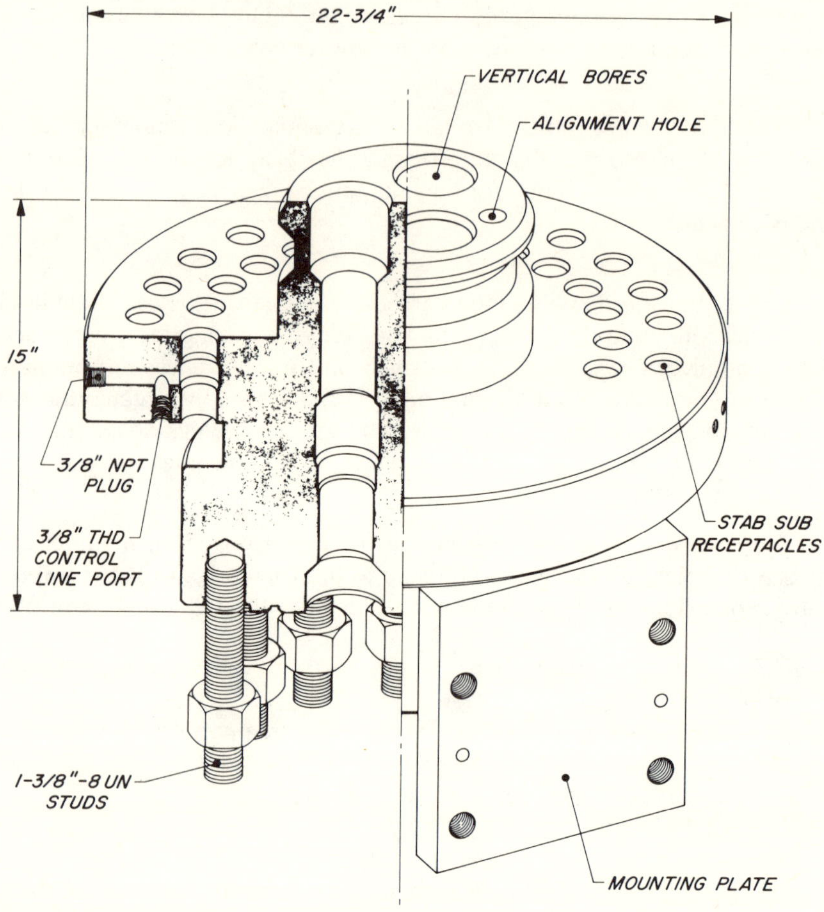

FIG. 11–10 Tree manifold. *(Courtesy Vetco)*

With the flow lines attached to the tree and the tree cap assembly installed, the operating pressure is routed through the flow-line connection and tree manifold to the sequence valve, where it is routed either back to the electrohydraulic control pod or to the individual tree function.

Tree Cap Assembly. When the well is completed and ready to be produced under direct hydraulic control from a remote location, the tree cap assembly is installed on the tree (Fig. 11–11). The tree cap protects and seals

FIG. 11–11 Tree cap assembly.

the vertical bores of the tree. It serves as a hydraulic control-line connector in conjunction with the tree mandrel. It also serves as a conveniently retrievable package for the various hydraulic control components housed within.

The tree cap utilizes an integral hydraulically actuated connector to latch onto the tree mandrel. A similar mandrel is affixed to the top side of the tree cap, which allows use of the tree running tool for running and retrieving operations.

Tree Control System

Typical of most diverless tree installations is the provision for controlling tree functions from either of two control points: (1) the surface vessel providing direct overhead control of the tree during installation of workover, and (2) the land-based or platform-mounted production station, from which remote control is maintained during production or maintenance.

The hydraulic control system on the installing or workover vessel provides the power as well as the controls for direct control during installation of the tubing hanger and tree, connection of the flow lines, and during all workover operations. For the satellite tree, this direct control during installation is attained through the running tool, which is connected to the tree manifold. After installation and with the tree cap in place, control from a remote location is attained through hydraulic and electrical umbilicals, which are an integral part of the flow line connection.

The satellite tree under discussion utilizes an electrohydraulic control system as the primary means of remote control with a sequenced hydraulic secondary or back-up system. The sequenced system can be actuated remotely through incremental changes in the hydraulic input pressure. In the primary operating mode, hydraulic fluid is directed to the electrohydraulic control system components at normal system operating pressure. At other predetermined operating pressures, the sequencing system directs the hydraulic operating fluid to manifolded tree functions. Remote service capability is provided through incorporating all tree-mounted components of the control system into retrievable modules.

Flow-line Connection System

Diver-Assist Flow-line Connection. The most common type of flow-line connection to a satellite tree in water depths up to 600 ft is a diver-assist flanged connection. For a second-end connection where the flow line is laid to the tree with a flanged end, there are two options for making the connection. First, the diver can maneuver the end of the line into position making it up to the tree with a misalignment union. Second, if the line is large or difficult to move, a jig can be used to measure for a flanged jumper spool to be fabricated on the surface and then lowered into position for diver makeup. Divers have use of underwater mechanical and hydraulic come-alongs and tools as well as lifting lines from the surface to assist in the flow-line flange connection.

Diverless Flow-line Connection Systems. The diverless flow-line connection system shown in Fig. 11–12 is designed for tree and flow-line installation as independent functional operations. The flow lines can be laid and attached to the wellhead structure before or after running the tree.

The system also has the versatility of allowing either a first-end or second-end connection. For first-end pull-in, the flow lines can enter from any angle in a vertical plane—from directly overhead to below horizontal. Second-end pull-in can also be accomplished with the flow line laid on the ocean floor at an angle of up to plus or minus 15° off centerline.

When the flow line pull-in tool is run, landed, and latched to the flow-line alignment structure, it incorporates all of the functions necessary for pulling the flow-line bundle to the tree and for precise positioning and final locking of the flow-line flange in the alignment structure. The flange is then in position for subsequent engagement with the tree flange and seal plate. This is accomplished with hydraulic positioning and locking mechanisms, which are part of the tree.

Tree Installation and Workover System

Tree installation and workover normally is carried out from a floating drilling vessel, since nearly all of the handling equipment required for these operations is utilized in the drilling operation and is already onboard. The major additional items required are a completion riser, tree running tool, and control skid.

The tree running tool is used for running and retrieving both the tree and the tree cap as individual units (Fig. 11–13). The tool is run on a completion riser or may be run on drill pipe or individual tubing strings in conjunction with a control hose bundle. The tool has an integral hydraulically actuated connector that latches onto the tree manifold or the top profile of the tree cap.

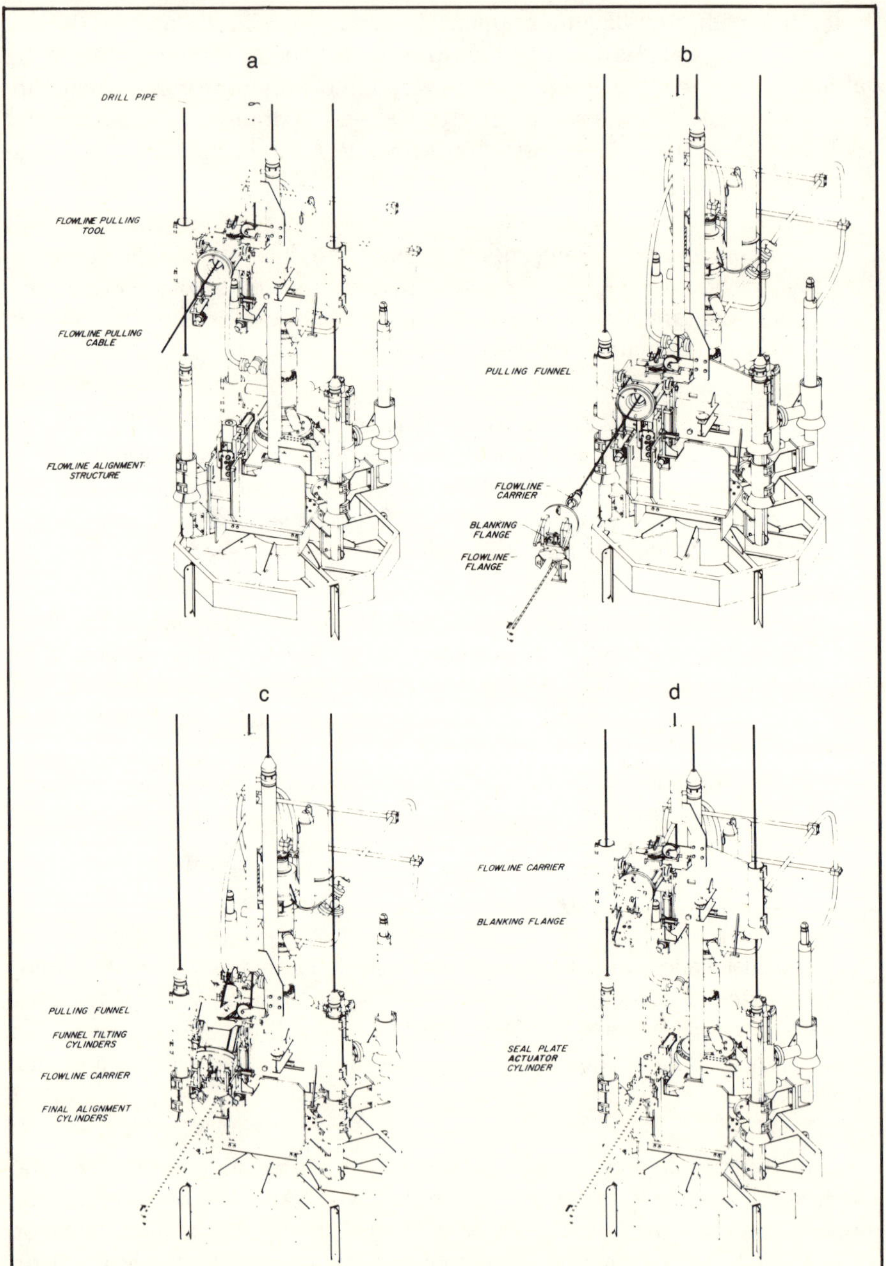

FIG. 11–12 Diverless flow line connection sequence.

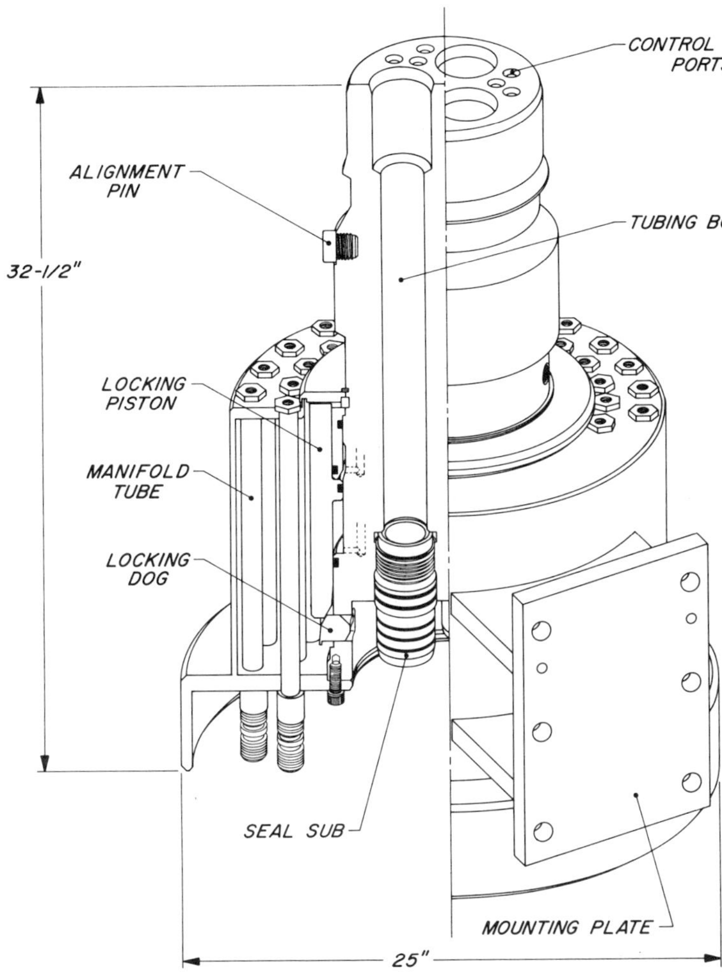

FIG. 11–13 Tree running tool.

The tree normally is placed on a test stump on the installing vessel and is operated through functional cycles and pressure-tested prior to being deployed. The tree is run on the completion riser over guidelines and is landed and locked to the wellhead. It is again functioned and pressure tested. All downhole work is completed prior to retrieving the running tool and completion riser. When the running string is retrieved, the well bore is secured by the master valves and the tree is ready for installing the tree cap assembly.

The tree cap assembly can be run in the same manner as the tree and is latched in place on top of the tree to seal off the vertical bores and complete the tree control package. For workover reentry, the tree cap is retrieved and the tree running tool with completion riser is installed to gain downhole access to the well from the surface.

DRY SATELLITE COMPLETIONS

A packaging technique to allow the use of surface-type Christmas trees and valves in a sea-floor application, dry subsea completions can be utilized for field development in the same ways as wet completions. Dry completions are made possible by using a pressure-resistant vessel that surrounds the well equipment and encloses it within a bubble of air at atmospheric pressure. This means workers can hook up, repair, and maintain the well equipment in a shirt-sleeve environment.

Workers are transported to and from the wellhead chamber by diving bell or submarine. The transport vehicles, which are a necessary component of the dry subsea wellhead system, must be able to lock on to a wellhead chamber, evacuating the water from the transfer trunk between the chamber and the transport vehicle, purging the atmosphere in the chamber, and pressurizing the chamber and transfer trunk with breathable air before workers can move from the vehicle into the chamber. They must also purge air from the chamber and refill it with inert gas before returning to the surface. The complexity of the equipment required to service a dry wellhead chamber safely has been a major drawback to general acceptance of this production method.

A one-atmosphere chamber, commonly known as a wellhead cellar, is designed to attach directly onto the casing head of the well with a hydraulic connector of the same configuration as a BOP stack connector. Installation typically is made by running the complete wellhead cellar, loaded with the assembled tree components, on a special drillpipe running tool from a floating drilling vessel. After coupling the cellar to the wellhead, the drilling rig moves offsite and the service capsule (a specially equipped diving bell) or a special submersible with dry transfer capability descends and couples to the wellhead cellar. Assembly and hookup of the tree and flow-line connections can then be completed by workers inside the protective chamber.

When multiple subsea wells are installed, they may be tied back individually to a fixed platform or template, or they may be routed to a subsea manifold center, also constructed as a one-atmosphere pressure vessel. This man-rated, ocean-floor habitat houses valves, piping, chokes, and controls

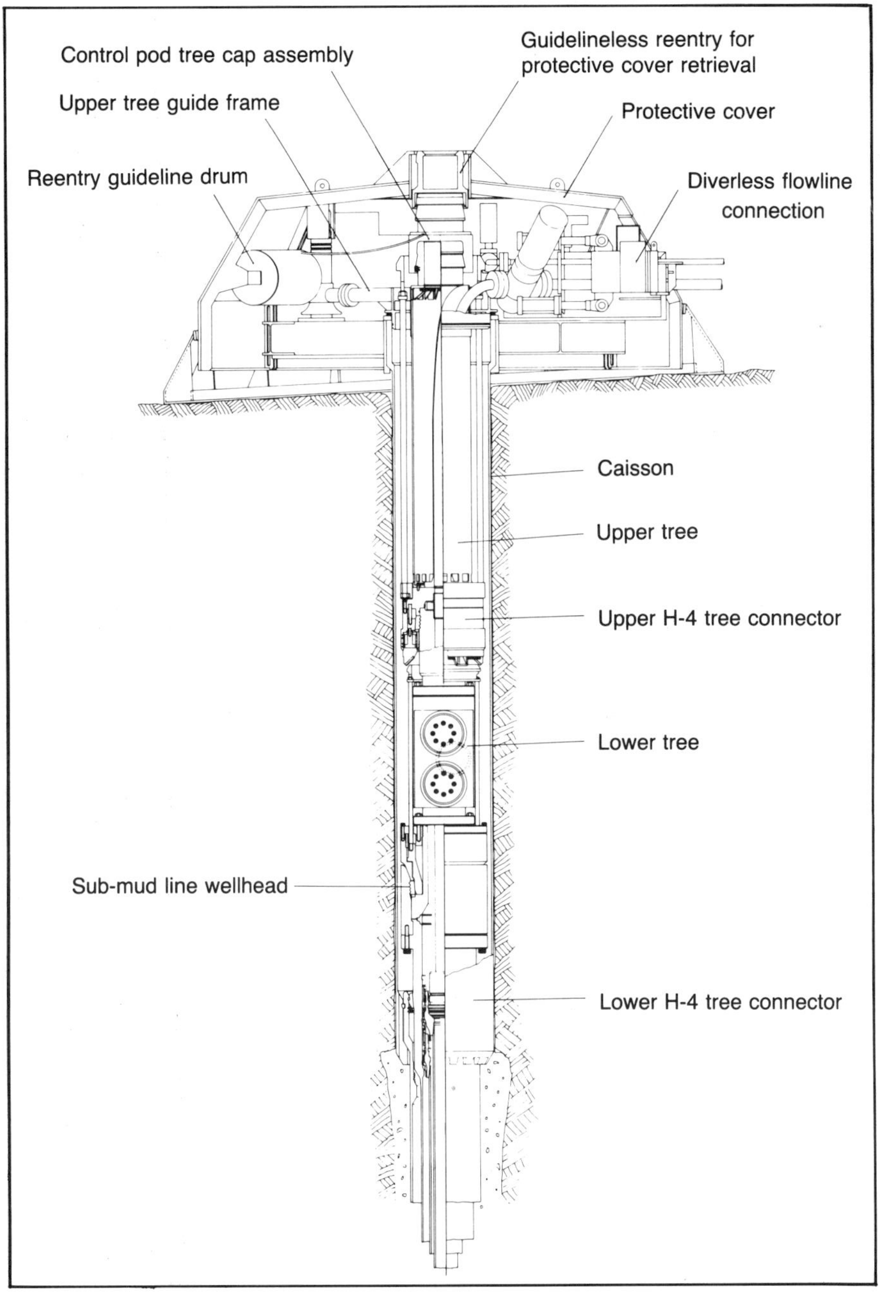

FIG. 11–14 Submudline completion system.

and commingles production from all wells. Then it pipes it to the surface for processing and sales, in addition to providing a central control point for all wells.

The basic components inside the subsea chambers are essentially identical to normal platform completion hardware since they, too, operate in the dry. Because a great deal of equipment must be installed in such a constricted space inside the pressure vessel, the packaging of the systems is critical. Naturally, the area taken up by items such as flanged connections and the acccessibility required for their alignment and installation puts space at a premium. For these reasons, special hardware items are used exclusively in dry subsea completion installations.

SUBMUDLINE COMPLETION SYSTEMS

Another type of satellite wet completion that, to date, has had very limited acceptance is the submudline or caisson completion system (Fig. 11–14). This system consists of placing most of the production tree in an extended conductor housing well below the ocean floor—out of the way of icebergs, anchors, and fishing nets. The upper part of the tree and the flow-line connection is housed in a low-profile protective cover.

FLEXIBLE RISER

Previous mention was made of early production of subsea satellite trees through flexible flow lines to a floating production platform or vessel. In this system, the flexible flow line is suspended from the anchored floating platform, and a surface connection is required should an emergency arise where release of the flow line and control line bundle is required.

The surface quick disconnect consists of a female connector, mounted permanently to the vessel's hull, and a male mandrel that has the production flow line, annulus line, control line, and electric cable risers attached. FSC gate valves above and below the connection contain the produced crude, should it become necessary to release the flow-line bundle in an emergency. A similar valve is located in the annulus line on the lower half of the connector for the same purpose. If the lines are released, they drop to the ocean floor. Retrieval is accomplished with guidelines attached by divers or submersibles.

IV
Production Control

C. Peterman *Hydril Co.*

12 Wellhead Control Systems

Systems for wellhead control must be approached in a generalized manner. Each oil or gas field usually has its own set of characteristics that must be recognized and adapted to the controls required.

Most offshore wells are drilled and developed from fixed platforms. Presently, about 2,400 wells per year are drilled and completed offshore. Each offshore operator has standards that are used for the control systems. These standards may be developed internally or they may be developed by outside consultants. They may also be developed through cooperation with one or more of the control system vendors that have emerged to service this portion of the offshore industry.

Therefore, the subject of control systems for offshore wells is complex, resulting from the need to tailor each system to the specific field requirements and to meet the specific design criteria furnished by the offshore operator, which may be developed internally, designed by consultants, or furnished by control vendors.

An additional offshore control system technology poses the greatest challenge for the control system designer: subsea wellhead controls. About 40 wells per year are being installed subsea. In subsea, one adds the elements of water depth, distance from the surface control location to the subsea wellhead, economics, and installation restrictions (buried control lines, wellhead covers).

PLATFORM WELLS

Each offshore well normally is equipped with some type of downhole valve and a group of surface valves known in the trade as the Christmas tree.

The control system design provides an interface between the well operator and the valves so they can be opened and closed under a prescribed set of conditions even if the operator is not present.

DOWNHOLE VALVES

Two basic types of valves are used downhole: flow velocity valves designed to close on predetermined production rates, and surface controlled subsurface safety valves (SCSSV) that open and close in response to surface control signals generated by the control system. Control systems are involved only in the latter category.

There are two basic categories of SCSSVs. The categories cover the control system design utilized from the standpoint of control pressure required and the number of control lines needed to correct the control panel (surface) to the downhole valve SCSSV. Commonly, these are designated as single-line and dual-line valves. They are also distinguished by whether the valves are tubing retrievable or wire-line set (Fig. 12–1).

SURFACE VALVES

The normal minimum complement of Christmas tree valves is a master valve, a wing valve, and a lubricator valve. The usual minimum control system requirement is to have a pneumatic or hydraulic operator on the wing valve that can respond to control system signals.

The master valve is in line with the tubing string. It is normally flanged to the tubing hanger upper surface. The lubricator valve is also located in line with the tubing run and is located on a flanged tee attached above the master valve. The wing valve is located on the other opening of the tee.

One common opening and closing technique for wellhead valves is to close the wing valve first, the master valve next, and the SCSSV last. The theory is that the wing valve can be replaced without killing the well and should be used to stop the flow. Routinely, the rest of the valves are closed only when the flow is stopped. This reserves the master valve and SCSSV for closing during emergencies when the preferred sequence cannot be followed.

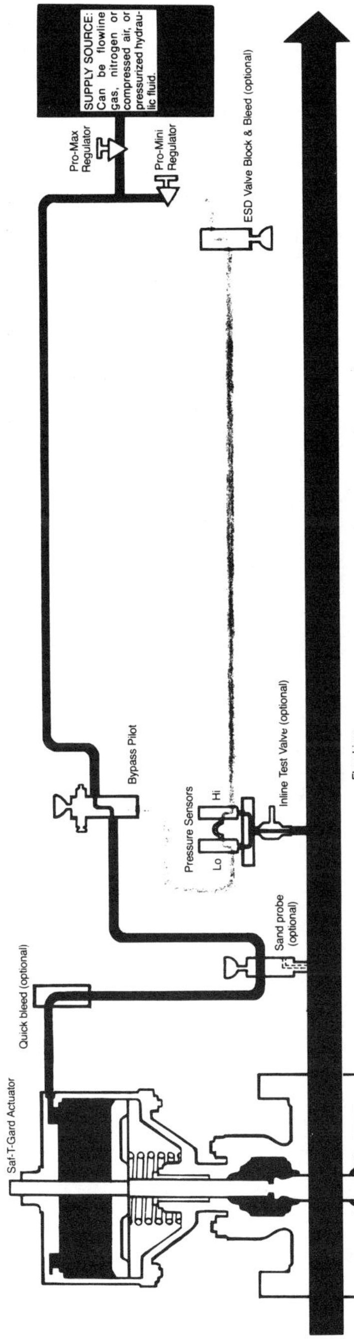

FIG. 12–1 Schematic of a typical W-K-M pneumatically operated safety system. *(Courtesy W-K-M)*

WELLHEAD VALVES

The main control system criteria are controlled by the type of power used to control the wellhead and SCSSV valves. Pneumatic supplies, e.g., wellhead gas, compressed air, or bottled gas from 60–150 psi, are often used to control the safety wing valve. These safety valves normally are equipped with springs to close the valve. The pneumatic supply opens the valve and compresses the spring. In this fail-safe or fail-closed design, the valve stays open as long as the pneumatic signal is present.

Hydraulic supplies of 1,500 to 10,000 psi are used for SCSSVs and may be used for safety valves to replace the pneumatic supply. The pressure required depends on the design of the valve operator.

For single-line SCSSV units, the control pressure must exceed the tubing pressure since one side of the valve operator is essentially exposed to tubing pressure in the single-line valve. For dual-line SCSSV units, the back side of the valve operator is exposed to a second control line pressure and is sealed from exposure to most if not all tubing pressure. This is called balanced design. In some SCSSVs, the downhole safety valve can lock open for wire-line operations by special, sequenced pressure applications to each of the dual control lines. Operating pressures ranging from 1,500 to 3,000 psi are normal for these valves.

The selection of wellhead valves and SCSSVs immediately establishes an important control system design parameter. Which type of power supply is required and what pressure and volume must be available are also important considerations.

INTERCONNECTIONS

This is not a critical item for Christmas tree valves. However, various types of materials are necessary for satisfactory operation in the offshore environment. Stainless-steel tubes and even plastic encapsulations over the tubes are used for this environment.

SCSSV applications, however, have a more complex interconnection. The control tubes must pass from the control panel to the wellhead, through the wellhead, and down the production tubing to the SCSSV. Numerous types of designs are used. Conventional setting depths for SCSSVs are 600 to 800 ft below the wellhead. A trend has been developing to increase this setting depth.

The wellhead equipment has various ways of passing a high-pressure signal through the wellhead to the bottom side of the tubing hanger. At this point a hydraulic tube (or tubes if a dual-line valve is used) is connected to the tubing hanger connections. Using base hydraulic tubes attached to the production tubing with bands is the minimum approach. A method of encapsulated hydraulic tubes using centralizer-type clamps to prevent crushing the tubes during installation is being used more often as a trade-off between low-cost connection links and trouble-free, expensive connection links.

The accepted standard, 316 stainless-steel tubes, is normally supplied in ¼-in. or ⅜-in. OD sizes. In some cases, operators are specifying more corrosion-resistant materials, such as monel or imonel, for severe services. The tubing is furnished in continuous lengths to eliminate the need for unions between the two end points.

CONTROLS

The controls involve the power supplies, the control valves, and the method of packaging the control logic and the safety sensors. Power supplies are determined by two factors: 1) What is required by the wellhead and SCSSV valves, and 2) What power supplies are available from the platform (electrical power for diving hydraulic pumps), air supplies, pressure, and volume. These are all subject to local conditions.

CONTROL VALVES AND PACKAGING

Normally, a manual panel can be manipulated by the platform production personnel (Fig. 12–2). It has labels and other instructions to guide these people in its correct operation. The control valves that furnish the signals directly to the wellhead valves can be panel mounted and manually operated. They can also be piloted valves that respond to signals from a manual panel control and are remotely controlled from various sensors. Combination valves respond to both manual and piloted operations. Other packing aspects are determined by space and environmental conditions.

Entire valve and hydraulic supply packages are often supplied. These are completely assembled and tested units contained in weatherproof housings with enclosed floor structures. The use of stainless steel for panels and other exposed surfaces is common. Electrical equipment is normally explosion proof to NEC Class I Division I (USA) or equivalent standards. Typically, all of the panels are enclosed in one unit of a size and structure that can be lifted from

FIG. 12–2 Control panel for a facility collection and separating oil and gas production. *(Courtesy W-K-M)*

a work boat and installed on the platform. In some cases, delivery takes place as the top works are being constructed and assembled.

Control logic is supplied in many forms. One of the most common is sequential pneumatics. In this design, timed sequences can be achieved by using metering devices and tanks to create time delays between events. These time delays let the wing valve close before other closing operations are permitted.

The use of electronic equipment has been very limited to date to achieve various control logic sequences. However, this technology is being evaluated more often for complex situations.

There is also a tendency to use logic to instigate well startup as well as shutdown. This ensures that the appropriate opening sequences and elapsed times are used to eliminate excess wear on the wellhead valves and SCSSVs.

In the simplest form, each well is equipped with an individual panel that may be mounted adjacent to the well or at a remote location. In some cases these controls are furnished with their own pneumatically powered hydraulic pumps and accumulators.

REMOTE SENSORS

The remote sensors continuously monitor pressures and flow conditions in the production line to provide advanced warning of dangerous conditions that are sufficient to close in a single well or all wells on a platform (Fig. 12–3).

HIGH-LOW PRESSURE SENSORS

This is a spring-loaded block and bleed valve that controls a pneumatic or hydraulic signal and monitors the pressure on the flow line downstream of the safety wiring valve. Normally, two units are provided: one for high pressure and the other for low pressure. The high-pressure unit detects unsafe buildup of the production line that might cause damage to downstream production equipment and lines. The low-pressure signal prevents loss of production through downstream leakage.

As seen from the location, once the wing valve closes, the high- and low-pressure pilots are denied their source of correct operating pressures. Bypass valving must be provided to let the well come back on production before the pilots can be placed in normal operation.

The pressure pilots are furnished with adjustable springs to change the pressure settings and production equipment condition. These adjustments are

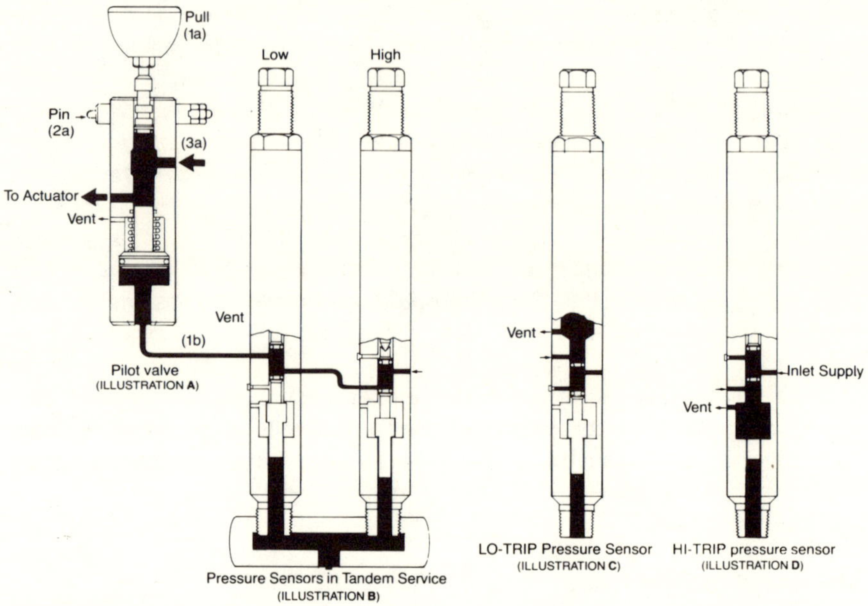

FIG. 12–3 Safety system operation under normal condtions. *(Courtesy W-K-M)*

available over relatively narrow ranges, and pilot valves or actuator adapter kits must be used to handle widely varying pressures.

SAND PROBE

These units are placed in flow lines to detect erosion caused by sand-laden production. The valve is equipped to detect flow line pressure applied to a pilot valve that occurs when the sand probe fails and lets the pressure reach the pilot valve. The probe must be replaced if it is the cause of a shutdown.

FIRE DETECTORS

These elements are used to detect abnormally high temperatures that in turn let the pilot pressure feeding the detector pilot valve bleed to the atmosphere. This creates a shutdown signal that causes the control system logic to respond and makes the wellheads shut in.

ESD (EMERGENCY SHUTDOWN) LINES

This is normally used as a control loop that is accessible from a number of points. It takes pneumatic pressure from a reliable source and passes through all of the ESD circuitry to corresponding logic units located on each control panel. The logic units respond to loss of ESD signals. Each control element in the line reacts to its own special input by blocking the supply side and bleeding the downstream side of the ESD line. ESD valves commonly are located at strategic positions on the platform, such as near the wellhead bay, by the boat landing, or on the helicopter deck. Other devices, such as production separators and dehydrators, are protected by high-low pilots in the ESD circuit.

The ESD line forms a continuous control link that can be actuated from any of the points mentioned above as a result of manual or automatic alarms. It is blocked upstream of the alarm point and bleeds the downstream circuitry, which, in turn, causes the wellhead control circuits to respond as planned to loss of ESD circuits.

IN SUMMARY

The wellhead controls are designed for a hierarchy of control levels. The first is the manual control. The second may be automatic sequences. The third is response to specific well conditions. The fourth is response to a general or ESD level. The last three are programmed into the individual panels. These responses are tailored to the specific platform requirements.

13 Subsea Production Control Systems

The primary objective of a subsea production control system is to open and close wellhead valves and SCSSVs. The main design considerations involve the location of the wellhead. The wellheads generally are located on the ocean floor and are separated from the control panel by very long distances.

Subsea completions have been used for the last 15 years. A steady increase in actual installations has been observed over the past 8 years. Initially, the controls for these systems were very simple and involved the use of a single hydraulic control line. This control line was connected on one end to all valves so they opened and closed to the same signal. On the other end, a hydraulic power supply and a manually operated control valve were provided. This approach worked reasonably well until a well did not respond properly to opening the valves. That brought about the need for obtaining information about valve positions and wellhead pressures.

The next generation of controls involved the adaptation of electric control cables, solenoid valves, and hydraulic supply lines. The electric cables provided easy access to control the solenoid valves and gave valve status and pressure transducer readouts. However, the hardware was adapted from off-the-shelf commercial components, and the application in the subsea environment caused difficulties. Exit electrical controls; enter simple hydraulics again.

However, the application of SCSSVs for subsea wellheads began to complicate the control systems because of the need for at least two control signals at two different control pressures. Also, the sequencing of valves for opening and closing was recognized as the best method for Christmas tree operations. Yet another design criteria began to emerge concerning opening and closing times for the wellhead valves. This all brought about the use of electrical control lines again, especially since the applications for subsea wellheads required distances of 5,000 ft or more from the platform. In fact, at one time a leading control system vendor took the position that multiplexed electro-hydraulic control systems were necessary at distances exceeding 5,000 to 6,000 ft.

There are several existing multiplexed EH control systems that are controlling subsea wellheads successfully. However, now the trend is away from multiplexed EH and toward direct hydraulic controls for wells completed as far as 15,000 to 20,000 ft from the control point. Long, continuous stainless-steel tube assemblies encapsulated in plastic jackets (hydraulic control umbilicals) have made this possible. The expansion of early designs of plastic reinforced under pressure were overcome by the use of steel tubes. Reportedly, presently available plastic-reinforced hydraulic hoses have reduced expansion characteristics. Some offshore operators are using electrical cables to connect wellhead and downhole sensors to displays on the hydraulic control cable panels furnished as part of the control system. These are used in addition to the hydraulic control umbilicals mentioned above.

Where does the offshore industry stand today? First, many control systems that are offered appear to do the same job. They can be classified as discrete hydraulic controls (direct or pilot operated), sequenced pilot operated hydraulic controls, and electrohydraulic controls (multiconductor cable direct control or multiplexed control).

Three production equipment factors must be considered in any of the systems. These are associated with production platform equipment, control linkage from the platform to the subsea wellhead, and wellhead control equipment. And there is a fourth factor to be considered: installation and maintenance (I&M).

The complexity of the I&M system varies, based on whether there is significant diver assistance. The I&M system controls all functions for the Christmas tree. The necessary controls to install the wellhead are usually included in the I&M system. These can involve actuating flow line connectors, wellhead connectors, and swab valves and pressure testing seals.

PRODUCTION CONTROL SYSTEM COMPARISONS

There is an ascending degree of complexity for each control system. There are also limits to the number of functions that can be controlled with each system (control capabilities). The selection process is a tradeoff between factors listed above, which eventually must be resolved in comparison to cost, complexity, and capability. These are influenced by the design parameters of each subsea wellhead candidate, which include the following:

Wellhead water depth
Distance from platform-based control to wellhead
Number of control functions
Status requirements
Support vessels available for installation and maintenance
Production rates
Technical competency of support personnel
Environmental factors (wind, waves, currents, temperature)
Desired response times for control actions

DISCRETE HYDRAULIC CONTROLS

Each control function is achieved by a separate hydraulic control line. The hydraulic supply furnishes high-pressure fluid to the inlet port of the control valve on a control panel located on a nearby fixed or floating platform.

Direct controls utilize a control line (tube) to transfer the hydraulic fluid directly into the subsea valve operator and to return the fluid in the operator back through the control valve and into the hydraulic power unit tank when it closes. All of the control tubes generally are assembled in a hydraulic control umbilical. This system is very simple and is the least complex on the list.

The closing and opening times for this type of system depend on hydraulic supply pressure, control valve size, control line size and length, elevation difference between the valve and the wellhead operator, and pressure-volume relationships for opening and closing the valve. There are complex relationships between all of these factors. For instance, if the operator has a satisfactory time response for a direct control system with the well located 5,000 ft from the platform and he wants to duplicate that operation (time response) for a well located 10,000 ft from the same control panel, he must increase the ID of the control line. All other factors remain constant. Evaluating the extension of this discrete direct hydraulic control from one distance to another involves only the increase of the control line size if the same open-close performance is to be retained.

Evaluating the discrete direct hydraulic control system, we find it relatively high on the cost side (large control lines required) and very low on the complexity side (one line-one control relationship). The capability of the system, in terms of opening and closing times, can be controlled by the size of the control line.

PRESENT TRENDS

The present trend in the use of subsea wellheads is to recover reserves that could not be reached from existing surface structures. These reserves are generally too small to warrant constructing platforms for more conventional production equipment. Another trend is to use subsea completions to accelerate the production from offshore reserves with production flowing to temporary floating production facilities.

In both cases, most offshore operators are using discrete, direct hydraulic controls. Distances of 3 miles from control panel to wellhead are served routinely with a combination of stainless-steel control lines and manual hydraulic control panels on the surface. Opening and closing times of 3 to 5 min are being accepted, so piloted hydraulic schemes are not required.

Piloted hydraulic systems have a control valve located at the wellhead. A small pilot line is used to shift the control valve. A hydraulic source, generally in the form of a hydropneumatic accumulator, is also located on the wellhead. The combination of piloted control valve and accumulator provides the capability to open and close the wellhead gate valves from 10 to 30 sec after the pilot valve shifts in response to a pilot signal applied to the pilot line.

Pilot valve shifting times depend on length, diameter, fluid viscosity, pilot pressure, and pilot valve operating characteristics. Water and soluble oil fluids used in ¼-in. OD stainless-steel lines can shift valves in less than 30 sec at distances up to 30,000 ft.

The economic benefits of using smaller pilot lines for this type of system are offset by pilot valve costs and the inevitable reduction in reliability by adding components to the system. This is not an important consideration if correct pilot valves are used with appropriate system designs. The reliability of the pilot valves is very high, and the resulting reliability of the combination pilot line and pilot valve is not noticeably changed when compared to a direct control approach.

Sequenced pilot valve systems have been introduced to minimize the expense of the hydraulic control lines linking the platform to the subsea wellhead. A variety of proprietary systems have been used. One of the main drawbacks is that the sequence must be programmed in advance. This means that all control requirements have to be predetermined. This looks good on the drawing board but is generally deficient in actual operations when irregularities in subsea well response occur. Most offshore oil and gas operators are presently using, or are considering the use of, direct or piloted discrete control systems rather than accepting the lower cost but less flexible and more complicated sequenced control approaches.

EH CONTROLS

The final category of control techniques involves the use of electric cables for control signals rather than the hydraulic lines. If the operator requires a complicated wellhead control (in excess of 8 to 12 functions), then the use of electrohydraulic controls may be necessary.

If the control distance exceeds 20,000 to 30,000 ft, the operator may also require the use of EH controls. If he feels he needs to measure pressure, temperature, and other wellhead values, the use of EH controls may also be warranted. Multiplexed techniques that eliminate the need to provide multiconductor cables are now proven. Typically, a control cable has conductors for signals (twisted, shielded pairs, or coax signal cables) and for electric power. Subsea cables generally have a substantial plastic jacket applied around the conductor package. This, in turn, is protected by an armor package of contrahelically applied steel wires. An outer thermal plastic jacket is often applied over the armor package.

Subsea electrical connectors are now available that can keep sea water away from the vital pin-and-socket connections. Some are even designed to connect and disconnect underwater. Another technique for making underwater connection is inductive couplers. Essentially, these are transformers cut in half with a primary coil on one side and a secondary coil on the other. The effectiveness of this connection lies in the fact that the coils are sealed from sea water and that the magnetic portions of the couplers are not changed functionally by the briny environment.

DIRECT DISCRETE HYDRAULIC CONTROLS

Platform equipment is comprised of a hydraulic power unit and a control panel for each subsea wellhead. The control panel has a manually actuated control valve for each wellhead control function to be operated during well production. Current practice is to provide a production master and wing valve, a downhole safety valve, and an annulus master and wing valve.

The hydraulic power unit contains a hydraulic fluid reservoir, air and/or electric motor-driven hydraulic pumps, accumulators, pressure regulating valves, relief valves, filters, and strainers. The unit normally operates in an automatic mode. When the pressure in the accumulators drops below a preset level, the pumps start automatically to recharge the accumulators.

When it is time to open a subsea wellhead valve, the required panel control valve is shifted to the open position. This valve position allows high-pressure fluid (controlled by a regulator valve) to flow through the valve, and into and through the control line. This hydraulic flow in turn pushes the valve piston on the subsea tree to the open position.

The fluid is supplied from an accumulator. The withdrawal of fluid from the accumulator causes the pressure to drop, which in turn causes the pump to start recharging the accumulators by taking fluid from the hydraulic tank. The pump stops when the pressure reaches a preset high point.

To close this wellhead gate valve, the panel control valve is placed in the closed position. This movement isolates the high-pressure source and lets the cylinder port drain back to the hydraulic fluid reservoir. The spring in the wellhead valve creates a force for the valve piston to push the fluid back through the control line and into the reservoir.

Petroleum-based hydraulic fluids are often used for these types of systems since no hydraulic fluid is discharged to the ocean during a normal control operation. However, care must be taken to make sure the viscosity of the fluid is suitable to achieve the opening and closing times required for the system.

As indicated above, the control linkage for this type of system is often a relatively large-diameter, stainless-steel tube. One tube is provided for each subsea wellhead control function. There are no pilot valves or accumulators located on the subsea tree. This simplicity is what attracts many operators.

DISCRETE HYDRAULIC CONTROLS WITH PILOT VALVES LOCATED ON THE SUBSEA WELLHEAD

Pilot valves are located on the subsea wellhead to reduce the opening and closing times associated with the direct control system.

In addition to the pilot valves, an accumulator normally is provided at the wellhead. This provides a common source of pressure for the pilot valves. The accumulator is recharged through an extra control line from the platform. Pressure control is achieved by feeding the platform end of the supply tube with a constant pressure source, normally in the range of 1,500 to 2,500 psi. Fluid will flow to the accumulator until the accumulator pressure equalizes with the pressure source.

With the accumulator located at the wellhead, the supply line length is shortened. Thus, the distance between the wellhead and the control panel is not a factor in determining how long it takes to open a wellhead valve. Of course, this depends on enough time being available to recharge the accumulator after a control action using accumulator fluid takes place.

The pilot valves are controlled using small-diameter pilot lines in the control umbilical. In this type of system, one pilot line is used per pilot valve required. A pilot signal is introduced to the pilot line by switching the platform control valve so high-pressure fluid flows into the pilot line. Only a small volume of fluid is required to shift each wellhead pilot valve, so the size of the pilot line usually can be small.

When the gate valve on the tree is to be closed, the pilot valve is allowed to shift by removing the open signal from the platform control panel. When the pilot valve shifts to a closed position, the actuator piston—powered by the spring—forces the fluid through the pilot valve and into the exhaust. Normally, the exhaust is open to sea water. This means that the hydraulic fluids needs to be compatible with the sea water. Special hydraulic fluids that are water-based systems with soluble oil are used. This also complicates the pumping unit on the platform. Now it must deal with two or three fluids that need to be mixed: water, soluble oil, and sometimes glycol to keep the hydraulic fluids from freezing.

The control lines in the control linkage umbilical are smaller. Many operators consider this a benefit since it is less expensive than the control required for a direct system. But there are more requirements for using additional pilot valves and accumulators and for using water-based, biodegradable fluids. This type of system has at least one more tube in the control bundle than the direct system that provides for the common hydraulic supply line.

The pilot valves generally are packaged in separate containers or pods for easy installation and maintenance. Although vendors offer many good designs, fundamentally there are two types available. These differ from each other in the hydraulic connections. Several face seals are used in a tapered male-female configuration. Since the hydraulic interfaces are tapered, the pod must be locked to the receptable during installation.

The second type of connection uses pin-and-socket connectors. Again, there are two styles. The first uses one set of seals. When this connection is pressured, a force is developed to push the pin and socket apart. A second style uses two seals to pressure balance the connection so the pin and receptacle are not required to be locked together. Then the pods are guided into location on the subsea wellhead using either divers or guide lines.

MULTIPLEXED EH CONTROLS

The main romance of a multiplexed EH control system is that only a very limited number of wires in the control cable is required for operation. In contrast, a typical electrical control system requires at least one separate conductor per control function or data point. When there are many controls and data readbacks required for a subsea wellhead control application, the use of a multiplexed EH control is warranted. Duplex systems normally are used. These require two wires for communication from the control panel to the subsea wellhead and two more wires for communications from the subsea wellhead to the control panel.

The electronics package required at the subsea location is known as an RTU, or a remote terminal unit. The RTU contains a modem, a multiplex unit, an overhead section, control units, analog receiver units, status input units, and an assortment of power supplies.

The modem is needed to transmit and receive signals. It is electrically matched to the communication conductors. The modem is designed to output variable-frequency signals representing the input signals from the multiplexer. The output frequencies are selected to reduce or eliminate electrical interference that can appear on the communication wires as a result of the proximity to the power conductors in the cable. The input signal is a series of DC signals, representing in serial form the message being transmitted. The modem also detects variable-frequency signals coming in on the common lines from the master unit on the platform and outputs the series of DC signals required by the multiplexer input.

The multiplexer receives the input from the modem in serial form, a time sequence of DC signals representing a string of I's and O's that form a message unit. The multiplexer outputs this message in parallel form. All parts of the message are available simultaneously at separate output points, one for each space in the message. The multiplexer also reverses and transforms a parallel message input to the required output serial form.

The parallel I/O (input/output) is connected to the overhead section. This contains certain logic elements, either hardwired or provided by microprocessor, to determine if the incoming message is for the RTU, if it is a logical message through interpretation of a certain part of the message added for security, and if commands are contained in the message unit.

The output of the overhead section usually is connected by a bus, a common set of wires connecting a number of I/O units. These I/O's are for the commands, status input, and transducer signals required for the subsea unit to operate.

The control or command signals output from the RTU normally are used to shift solenoid valves. Solenoid valves form the electrohydraulic interface between the control system and the subsea wellhead gate valve operators. The solenoid replaces the pilot operator described previously for pilot-operated discrete hydraulic controls. Normally, the hydraulic power is furnished in the same way as piloted hydraulic systems.

PLATFORM UNIT

The platform unit required for the multiplexed EH system has many of the same features mentioned in the RTU. The interface between the operator and the system is a control panel. This panel has push-button switches that initiate commands to the RTU. The signal created by the switch closure forms an input signal to the overhead section. The overhead section creates a parallel output signal identifying the RTU, the command indicated by the switch closure, and a section of code for message security. This, in turn, is presented to the multiplexer and the modem and is then transmitted to the RTU. Return messages from the RTU are received, demultiplexed, and interpreted, and output is recorded from output logic units in the panel electronics on lighted displays on the control panel.

A well-designed system will provide a number of features to prevent accidental actuation of buttons on the panels. It will also have backup power supplies so occasional losses of platform power will not mean losing control of the subsea unit.

The most up-to-date systems provide fail-safe subsea unit designs so momentary loss of communications or power between the platform and subsea wellhead will not cause unnecessary production stops. The subsea controls in this case respond to loss of hydraulic signals to fail-safe as required. The fail-safe hydraulic signal is created at the platform hydraulic power unit and can respond to control-panel commands or to automatic platform alarm signals that may require subsea well shutin.

INSTALLATION AND WORKOVER EQUIPMENT

As the name implies, this portion of the control system is used for installation and whenever well workover is required. Common practice at this time is to use hydraulic controls for this, regardless if the production control system is discrete hydraulic or electrohydraulic.

A multihose workover umbilical is provided with one hose for each control function required. Normally, all wellhead functions are controlled by this system, including operation of swab valves, wellhead connectors, and flow line connections. Provisions are often made for direct control rather than a piloted valve at the subsea wellhead. The surface equipment involves a reel to store the hose umbilical, a control panel, and a hydraulic supply unit. Clamps are sometimes provided to attach the hose bundle to the workover riser, which is normally used with a subsea system.

Offshore Oil-Field Diving Operations and Equipment

Larry Cushman *The Ocean Corporation*

14

Commercial Diving

Commercial diving is a method of transportation that lets people perform tasks in the water or on the ocean floor. Today there are three basic ways people can work underwater. The first and most common is pressurized or hyperbaric diving. With this method the diver goes directly into the water to perform the work and is continuously subjected to the changing pressures of depth.

The second method of diving is unpressurized or one-atmosphere diving, in which the person goes underwater in a submarine, an unpressurized sealed diving bell, an armored one-atmosphere diving suit, or some other device that can withstand external depth pressure. The advantage of a one-atmosphere diving system, such as a submarine, is that the occupants are protected from the conditions inherent in a high-pressure underwater environment. They are not subject to the physiological changes caused by increased pressure, such as inert gas narcosis or decompression sickness. However, a person enclosed in a protective one-atmosphere diving system cannot perform certain types of work as effectively as a diver who is outside in the underwater environment.

The third method of diving, which is relatively new, is by means of unmanned remotely controlled submersible vehicles. These devices, sometimes called remotely operated vehicles (ROVs), underwater tethered submersibles (UTSs), or remote-controlled vehicles (RCVs), are operated by topside personnel with television systems and joy stick controls that operate the self-propelled underwater vehicles through long electronic control and TV umbilicals. ROV units can carry underwater instrument packages and are usually equipped with surface-operated electrohydraulic manipulator devices that are operated from the surface. They are often used for NDT (nondestructive

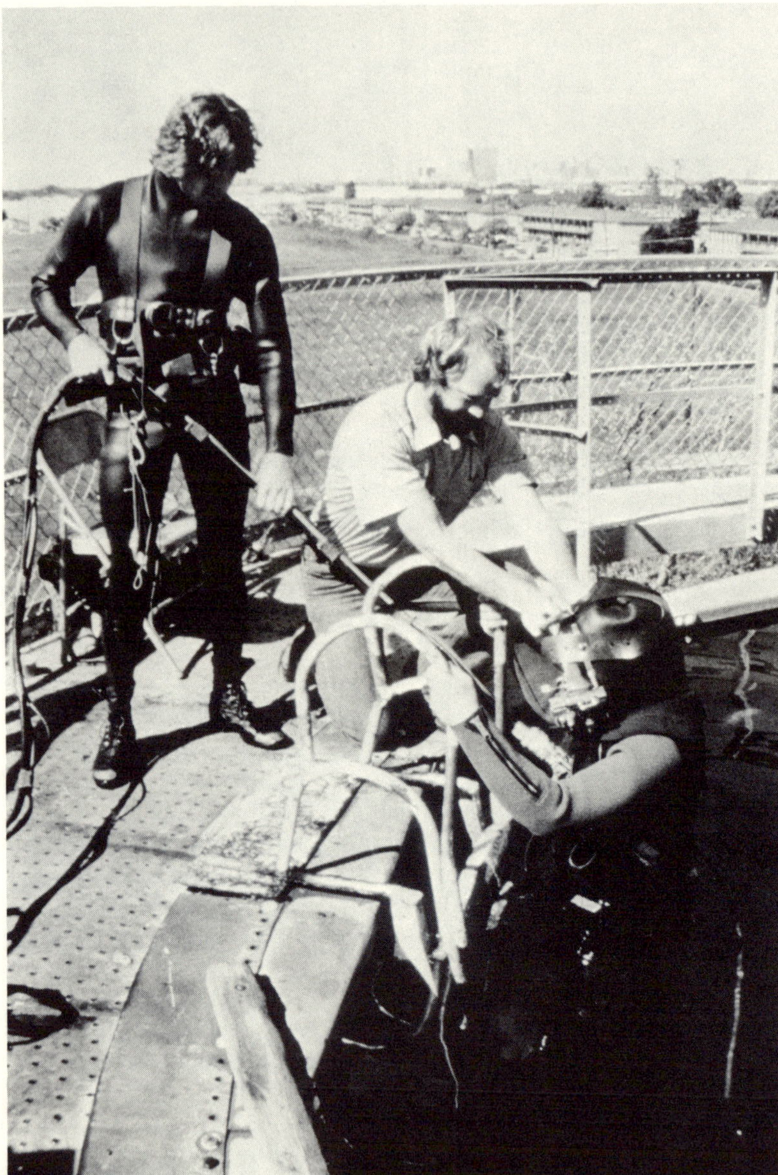

FIG. 14–1 A commercial diving instructor checks out a student equipped with lightweight surface-supplied air diving gear prior to a training dive in a test tank. *(Courtesy The Ocean Corporation)*

testing) underwater visual and TV inspections and light work where manned diving is impractical.

Today, the most effective underwater work is performed by divers in the water, in spite of the physiological problems that can sometimes result from hyperbaric exposure (Fig. 14–1). One-atmosphere and ROV-type diving systems generally complement conventional diving services and perform ultra-deep diving tasks, but no vehicle has yet been developed that has the overall underwater work capability and adaptability of the well-trained, experienced commercial diver.

15

History of Diving

The accumulation of seashell artifacts at prehistoric living sites indicates food was taken from the sea by divers long before recorded history. The earliest records are of Cretan sponge divers (3000 BC) and Chinese pearl divers (2200 BC). Military divers were used during the Trojan War (ca. 1194 BC) and are mentioned in Homer's *Iliad* (pre-700 BC). Alexander the Great deployed frogmen against the defenses of Tyre in 333 BC and was supposed to have descended in an open-bottom diving bell himself. Aristotle writes of the diving bell in the 4th century BC. Prior to this time, all diving was probably done by breathholding to depths not exceeding 60 or 70 ft.

The open diving bell was the dominant diving apparatus for the next 22 centuries. In the late 1600s the bell was refined, and in 1691 a large, sophisticated open bell was patented by Edmond Halley. This bell was ventilated by lowering barrels of fresh air. Dives were made with Halley's bell to at least 60 ft for as long as 90 min, and the divers made swimming excursions from the bell to adjacent work areas while holding their breath.

By 1770 the elementary hand-operated air compressor provided the next major advancement in diving. The compressor enabled LeHavre (1774) to develop a moderately successful helmet-hose diving apparatus. Surface-supplied compressed-air diving became the predominant diving technique by 1800 and was to hold a virtually unchallenged position from then on. In 1819 Augustus Siebe introduced the open-helmet-with-shirt system. This was the forerunner to his closed-dress diving system, which was developed and perfected to salvage the HMS *Royal George* in 1837. The Siebe closed-dress and hardhat helmet system was the primary diving apparatus for most working divers from 1837 until the early 1960s.

From 1819 to the present, progress in diving was dependent on two factors: improving the mechanical air compressor and studying hyperbaric physiology. The compressor improved rapidly during and following the industrial revolution; however, the study of diving physiology was slower to progress. In 1878 physiologist Paul Bert started to untangle the complexities of nitrogen absorption/elimination, or decompression sickness (the bends). The first hyperbaric chamber used to treat decompression sickness was installed to support caisson workers constructing the first Hudson River Tunnel in New York in 1893. In 1907, using much of Paul Bert's work, John S. Haldane produced the first decompression tables for divers.

In 1835 Condert published the design of a free-flow, self-contained underwater breathing apparatus (scuba) that consisted of a helmet, a flexible dry suit, and a compressed-air reservoir fitted around the diver's waist. This unique design had significant influence on the development of future diving apparatus. In 1865 Rouquayrol developed a demand regulator breathing system. Although this unit was surface-supplied by an air hose, it had considerable influence on the development of scuba. In 1878, Fleuss and Davis designed the closed-circuit oxygen scuba, which utilized a chemical absorbent to remove carbon dioxide. This was the beginning of a long line of closed-circuit oxygen scuba devices that led to the eventual development of the semiclosed-circuit, mixed-gas scuba by Lambertsen. Yves le Prieur, in 1924, introduced a manually valved, self-contained, compressed-air breathing apparatus. In 1942, Emile Gagnan and Jacques Cousteau developed the demand-type scuba regulator, which became the basic compressed-air scuba diving system design used throughout the world today.

The origination date of the U.S. Navy diving program is not actually known; however, official records indicate that George Stillson began developing the Navy's program around 1912. The F-4 submarine disaster of 1915, which somewhat paralleled the more recent *Thresher* submarine incident in terms of government and public reaction, stimulated considerable general interest in diving. The first U.S. Navy diving school was opened in 1915, and the Navy's experimental diving unit (EDU) was originated in 1927. Navy mixed-gas, helium-oxygen diving experiments began in the 1930s, and mixed gas was used extensively in salvaging the submarine *Squalus* in 1939. During World War II the advantages of military combat diving became evident, and in 1943 the famous USN Underwater Demolition Team (UDT) was formed.

Experiments in living in an underwater habitat on the sea bed became prevalent in the early 1960s. Ed Link (USA), the inventor of the famous Link aircraft trainer, put a diver underwater for 24 hr at 200 ft during his man-in-the-sea project in 1962. Jacques Cousteau (France) conducted Conshelf I (69 hr at 35 ft) later in the same year. In 1964 the first U.S. Navy underwater habitat living experiment, *Sealab I,* was conducted off Bermuda. Four divers stayed nine days at a depth of 193 ft. *Sealab II* and other projects followed as part of an ongoing man-in-the-sea research effort. Concurrently, Cousteau continued to conduct his Conshelf series of underwater living and work programs, which culminated with a successful 6-diver, 23-day, 328-ft saturation habitat submergence in 1965.

In 1966 some of the California-based commercial diving companies—including Ocean Systems, Divcon, and Cal Dive—began 300–650-ft bell diving and saturation diving experiments that were designed to show the oil companies operating offshore California that deep working dives were practical and safe. The commercial diving companies' concept of saturation diving, however, was different than that of Link, Cousteau, and the U.S. Navy. Rather than put the diver's habitat or living chamber on the sea bed, they put it on a surface vessel where diver life support and logistics were much easier to control. The working divers were lowered in a closed diving bell to the underwater work site and were returned to the surface-situated pressurized habitat chamber when their work sequence was finished. In the late 1960s, while the various government-financed and military sea bed habitat experiments were getting considerable international publicity, the privately owned U.S. diving companies were quietly perfecting their closed-bell diving procedures and were performing the first commercial deep-water and saturation diving work ever done.

The decade of man-in-the-sea experimentation by the aerospace, research, and military organizations began to slump in the late 1960s. At that time it became clear that many research diving programs would not result in immediate financial returns for the companies that supported them. Much of the scientific underwater experimentation was suspended, many of the underwater research divisions of the large U.S. aerospace companies were closed down, and most of the more than 70 submersible vehicles built by these companies during the 1960s were put into storage, sold, or scrapped. However, the practical development activities in oil-related offshore commercial diving continued to increase.

The demand for underwater services in the offshore oil industry introduced a new era in diving. The commercial diving service companies that consistently demonstrated their ability to perform safe, useful work underwater began to emerge as important contributors in the dynamic business of recovering oil from the sea. Today almost all of the deep, long-term, and most complex underwater work going on around the world is being done in support of the international offshore oil industry.

16

International Offshore Oil-Field Diving

Commercial offshore oil-field diving is by far the largest and most sophisticated diving activity in the world today. Diving service company revenues in 1981 were about \$500–600 million worldwide, about 60% of which was produced from non-U.S. operations. The growth of the business during the last eight years has averaged about 10–15% per year. The overall growth rate of the business in the near future is predicted to continue increasing, but at a somewhat slower rate. An exception is the rapidly growing international offshore inspection, repair, and maintenance diving business. This activity is increasing because some of the older offshore structures around the world are approaching the end of their design lives but will be maintained if possible because of the increased value of oil. The most practical and cost-effective way to inspect and repair these offshore oil production facilities today is to utilize the various commercial diving work capabilities.

Offshore oil-field diving can be related directly to each of the three major sequential offshore oil-industry activities it supports. The first offshore oil activity (after seismic survey, which does not require diving services) is exploratory drilling. The second activity is offshore construction (i.e., installing offshore oil production facilities such as production platforms, offshore loading buoys, subsea completions, and subsea pipelines). The third activity is maintenance of previously installed offshore production structures and facilities.

EXPLORATORY DRILLING VESSEL SUPPORT (RIG DIVING)

There are two primary types of exploratory drilling vessels: jackups and floaters. Each requires a different type of diving support service. Jackup drilling rigs are generally small units with limited deck space that operate in water depths of less than 350 ft and are usually close to shore. Most of the drilling equipment (such as the blowout preventer) is on the rig, not the sea bed, and can be serviced by topside personnel. For these reasons, most diving support work on jackups around the world involves relatively simple underwater tasks that usually can be done by what divers refer to as *callout diving*. That is, the diving crews and their shallow-water diving equipment remain on shore until needed. The divers mobilize their equipment onto a small support boat from which the diving is done (or sometimes onto the rig itself), perform the work, and demobilize their equipment back to their shore support base until called out for the next job.

Floaters are drillships and semisubmersible drilling vessels that normally operate in water depths greater than 200 ft. Some of the drilling equipment is situated on the sea bed and can only be serviced by remote control or, if this fails, by divers. Semis are designed for use in rough-water areas. Because floaters generally operate farther from shore and in deeper water than jackups, the diving systems are semipermanently installed onboard, and they are much larger and more complex than shallow diving systems. The divers live onboard the vessels and work the same shifts and rotation schedules as the drilling crews. They dive when needed, using bounce-bell diving procedures and equipment on vessels operating in the 200–400-ft depth range, and partial- or full-saturation diving procedures and systems on vessels operating in the 300–800-ft depth range, depending on the duration of the underwater task required.

Exploratory drilling vessels are designed to operate without the need to use divers. However, it has been demonstrated that diving services can reduce rig downtime significantly, especially on floaters operating in depths greater than 200 ft. The divers onboard floaters are, in a sense, emergency personnel who help minimize drilling down time. They also provide important underwater observation and inspection services when the rig or vessel begins and completes the drilling process.

The most common diver tasks on offshore drilling rigs are visual and television observation and inspection work. Guide wire replacement and debris removal are other common tasks. Sometimes a diver will be needed to change an A–X ring, replace hydraulic lines, or perform other relatively simple mechanical tasks that can eliminate the need to bring the subsea equipment to the surface for repairs. Divers can also help recover subsea equipment when no other method is possible.

Most diving tasks from a drilling rig can be accomplished within 30 min total bottom time using bounce-bell diving procedures. A study made by Oceaneering International a few years ago of over 3,000 working dives on floaters substantiates this. The study showed that more than 80% of all working dives were shorter than a half hour, and 99% were completed in less than one hour. An earlier Oceaneering study of a drillship operating in the Irish Sea showed a total of 222 dives averaging 16 min each.

Drilling vessel tasks that require longer bottom times can be accomplished with back-to-back bounce-diving techniques using a second deck decompression chamber and additional divers. Saturation diving can be used effectively from 200 ft but would probably be justified only if unusually long bottom times were anticipated.

The maximum safe limit for bounce-bell diving is not easy to define. Bounce dives have been made from drilling rigs working in more than 800 ft of water. However, for all practical purposes the maximum effective bounce diving limit today appears to be around 500 ft. Wise operators provide for at least minimum saturation capability (using what are now being called minisat systems) once their rigs get into water depths over 300 ft. The requirement to use saturation diving procedures in the 300–600-ft depth range depends on whether the diving crew can accomplish the tasks within the bounce diving time limits.

For example, a job at 550 ft that cannot be finished in time will force the crew to saturate in order to complete the task. After this dive the deck decompression facilities will be occupied for a five- or six-day period, and during that time no other diving can be conducted. On the other hand, one might deliberately saturate the diving crew if the job is the first of an anticipated series of tasks (such as might occur when working on a damaged BOP stack). The divers would then be retained in saturation at the shallowest possible storage depth and would be able to make continuous no-decompression excursions from the storage depth to the bottom for whatever time periods necessary to complete the work.

Once depths beyond 400 or 500 ft are reached, there is probably no alternative to saturation diving. The use of full-saturation diving systems and properly qualified personnel is mandatory.

OFFSHORE CONSTRUCTION SUPPORT DIVING

Construction diving is the catchall term divers use to refer to underwater work associated with installing offshore oil-production equipment such as

FIG. 16–1 A rig diver works on an underwater wellhead offshore of Southeast Asia. *(Courtesy Solus Ocean Systems)*

production platforms, offshore loading buoys (e.g., SBMs, CALMs, etc.), subsea completions, underwater pipelines, and risers. Diving is an integral, preplanned part of most of these operations. For this reason and because most production system installations today require extensive underwater work in the 200–600-ft depth range, full-saturation diving equipment and procedures generally are used.

The technology of saturation diving originally was developed to provide a more reasonable ratio of bottom working time to decompression time than is possible with surface-to-surface or bounce diving procedures. For any given depth there is a bottom time beyond which the inert gas the divers are breathing effectively saturates their body tissues. At this point their decompression becomes a function of depth only. Saturation diving differs from bounce diving in that decompression time does not increase when bottom times are extended, but the decompression requirement at the end of the dive series will be longer.

For example, it would take about four days to return a diver to surface pressure after a short series of working dives at 400 ft. However, the dive series could go on indefinitely at 400 ft without increasing the four-day decompression time requirement. In construction diving, most of which requires long work times in relatively shallow water, the longer time required for saturation decompression is a worthwhile price to pay for the capability to perform continuous underwater work with the same divers for a period of days or even weeks.

Construction diving spreads are made up of large multiple-deck decompression or living chamber complexes, one or more diving bells, and sophisticated environmental control systems designed to support as many as twelve divers living in saturation for extended periods. The work is carried on around the clock by teams of divers who continually rotate to and from the underwater work site in closed diving bells. Construction divers on major offshore projects may remain under pressure living in the deck chamber complex and diving in rotation for as long as 20 or 30 days.

Construction diving tasks are varied, ranging from observation and TV inspection to heavy underwater rigging and hyperbaric dry-habitat underwater pipeline welding. Most construction divers have progressed from shallow surface-supplied diving, mixed-gas diving, and rig diving to the offshore construction end of the business. They are generally the most experienced and capable of all commercial divers. Consequently, and because they often spend long periods in saturation, they are also the highest paid of all divers. Most construction divers work on a day-rate basis and are paid saturation premimum payments in addition to their daily wages. They generally work a complete offshore project during the summer construction season without taking time off except during periods of weather down time.

Almost all offshore construction diving work is performed from large derrick or utility barges owned by offshore construction companies, from specialty diving vessels owned by the larger international diving companies, or from the new semisubmersible construction vessels owned by the oil companies operating in the North Sea. Specialty diving vessels are usually pipe carrier-class supply vessels or other converted 200–300-ft LOA (length overall) vessels that have been modified to carry large construction diving spreads (50–150-ton cranes) and, in some cases, to support diver lock-out submarines. Some of the newer specialty diving vessels in the North Sea are purpose-built vessels designed specificially to support offshore construction operations. These specialty diving vessels are being used in the U.S. Gulf of Mexico and Latin America as well as in the North Sea and will probably soon be used in all major offshore construction areas of the world.

In 1981 construction diving represented about 60% of the combined revenues of the international commercial offshore diving service companies. This part of the diving business has been forecast to grow at a rate of about 7–9% annually over the next few years.

MAINTENANCE SUPPORT DIVING

From the standpoint of the commercial diving service companies, offshore oil-field maintenance means inspection/nondestructive testing (NDT) diving and the underwater repair work that often results from the inspections. In most cases the diving company that finds a problem is best situated to perform the necessary repairs. The underwater repair work requirement often is much more extensive than the initial inspection effort.

The requirement for underwater inspection and repair varies considerably worldwide, but most observers agree that the work requirement is increasing faster than any other segment of the offshore oil-field diving business. In the Gulf of Mexico, there is continuous underwater inspection/repair activity, mainly because the oil companies are concerned about the condition of some of the older production structures (Fig. 16–2). The average age of the almost 3,500 offshore structures in the area is about 12 years. Some are over 25 years old. A similar situation exists in some countries in South America, such as Venezuela and Peru. In addition, the increasing value of oil has made it

FIG. 16–2 A diver performs a platform wall thickness inspection using an underwater ultrasonic metal thickness detector. *(Courtesy Solus Ocean Systems)*

economically attractive for oil companies to continue producing from many older reservoirs that might have been considered marginal a short time ago. The desire to increase U.S. and Latin American domestic production to reduce dependence on foreign oil imports has created an even stronger interest in keeping the older offshore production facilities in good working order.

In the Norwegian and British sectors of the North Sea, all offshore oil-field structures, including newly installed facilities, are now inspected on a regular basis. This is because (1) many structures are standing in deep water in one of the most hostile sea environments in the world, (2) some of these structures are relatively new designs (e.g., concrete production platforms), and (3) government regulatory agencies in these countries are much more heavily involved in the offshore oil business than they are in the U.S. For these reasons, the underwater inspection activity in the North Sea is currently considered the most sophisticated in the world.

There is a considerable amount of offshore surface and underwater inspection and repair work going on in the Middle East because of the large number of older steel structures and the unique environmental conditions (i.e., very warm high-oxygen, high-salt-content water). But bell diving procedures and equipment usually are not required because many of the structures are in shallow water depths. Most of the diving work is done using surface-supplied air and mixed-gas diving techniques.

Most underwater inspection diving is done from appropriately equipped vessels positioned alongside the structures, over the pipelines, or from the platforms themselves. Saturation diving techniques are used when long-duration work times in deeper waters are required. Remotely operated vehicles (ROVs) play an important part in inspection diving, particularly in situations where only TV inspection is needed.

The most common underwater inspection tasks include visual and TV observation (to assess general structural conditions, marine growth, and sea bed scouring), marine growth removal and structure cleaning (using water blasters, grinders, needle guns, and hand tools), nondestructive testing (such as ultrasonic flaw detection), weld inspections (using magnetic particle crack detection techniques), and cathodic protection system inspections using bathycorrometers (underwater voltmeters). Some newer underwater inspection techniques, such as acoustic holography and acoustic emission stress measurement, are being developed for offshore use with varying degrees of operational success. Practical methods for inspecting large concrete structures underwater are also being developed in the North Sea.

Diving repair work related to offshore maintenance is similar to offshore construction support diving. All of the same diving equipment and capabilities are used, including hyperbaric dry-habitat welding. In addition, wet-stick welding is commonly used to make steel structure repairs underwater. Recent technical advances in dry-box underwater welding by several diving companies may displace wet welding as the primary underwater structural repair method.

17 Physiological Constraints in Diving

All diving may be categorized broadly as air diving or mixed-gas diving. The former, in which the diver breathes air, has a physiological limit dictated mainly by the narcotic or anesthetic effects of the nitrogen the diver is breathing at depth. A maximum depth of about 190 ft generally is accepted throughout the industry for surface-supplied dives using air. Below this depth all or part of the nitrogen must be replaced by another, less-dense inert breathable gas, usually helium, that has no apparent narcotic effect on the diver in depths to at least 1,500 ft. Almost all offshore oil-field diving deeper than 200 ft, whether it is surface-supplied, bell diving, or saturation diving, is done using a mixed-gas (helium-oxygen) breathing supply to avoid the narcosis problem.

DECOMPRESSION

Divers must decompress after a dive according to a predetermined ascent schedule so the inert gases (i.e., nitrogen or helium) that were absorbed into their body tissues during respiration under pressure will be released slowly without causing the bubble formation that would result in decompression sickness or bends. The quantity of inert gas absorbed during a dive is a function of pressure (i.e., depth) and duration of exposure (i.e., bottom time).

Several operational techniques minimize decompression times so diving equipment and personnel can be used to maximum efficiency. One of the most common techniques is short-duration or bounce diving. In bounce diving, the divers are pressurized for a short time (usually less than 1 hr) so they can be decompressed as quickly as possible. Their bodies have absorbed a relatively small amount of inert gas due to the short exposure to the increased pressure gradient while they are on the bottom, so their decompression can be completed in a matter of hours.

Today's knowledge of decompression requirements for the different depth levels is based on an extensive laboratory and field history of the time it takes to decompress the various types of body tissues without inducing decompression sickness. For example, a reasonable maximum limit for conventional bounce diving today is about 30 min at a depth of 400–500 ft. There has been little actual diving beyond that depth.

All decompression tables are based on theoretical calculations that have been followed by simulated (dry-chamber) dives. When these tests prove adequate, the tables are tested in wet chambers and under field operating conditions, and modifications are made based upon the results. So far, there has been insufficient operational use of bounce diving tables beyond the 600-ft level to provide the empirical data necessary to ensure the deeper bounce diving decompression schedules will eliminate the possibility of decompression sickness.

Another operational technique is saturation diving, introduced to the offshore oil-field diving industry in 1966. At any given depth there is a bottom time beyond which the inert gas saturates the body tissues and gas is no longer absorbed. At this point the decompression time requirement does not increase, so the diver can remain there indefinitely. However, the saturation decompression requirement under these circumstances is much longer than the decompression for bounce diving. The rule of thumb for estimating saturation decompression using existing tables is about one day of decompression time for each 100 ft of depth. This long decompression time may seem impractical; but any underwater job that requires extended work times, even in shallow water (such as most offshore construction diving projects), can be accomplished most efficiently and at less expense by using saturation diving procedures.

Theoretically, the body does not become completely saturated at any pressure until the diver has remained at that point for about 24 hr. In practice, however, saturation decompression is required whenever the diver stays underwater at any depth for a length of time beyond which he can be practically decompressed using conventional bounce diving or surface-to-surface decompression tables.

THE COMPRESSION BARRIER

Bounce diving deeper than 500–600 ft is complicated by what has been described as the compression barrier. Divers' bodies begin to absorb the inert gas they are breathing as soon as the pressure increases. The amounts ab-

FIG. 17–1 A 1,000-ft rated split lock saturation diving system installed aboard a Zapata semisubmersible drilling rig. *(Courtesy Oceaneering International)*

sorbed during descent can vary considerably due to descent delays, so they are unpredictable. For this reason, bottom time starts when pressurization (descent) begins, not when the diver actually reaches the bottom. Since the dive must be limited to a specific overall time, any reduction in descent time leaves more time for useful work on the bottom. For example, divers usually can be pressurized down to a depth of 500 ft in about 10 min. If the overall diver time limit is 30 min, the divers will have at least 20 min of useful work time on the bottom before they must begin their ascent and decompression schedule. If the divers could pressurize (descend) to depths beyond 600 ft at a similar rate, there would be more work time in deep water without extending the decompression requirement.

Rapid pressurization to depths beyond 500 ft can cause a problem known as High-Pressure Nervous Syndrome (HPNS). The HPNS mechanism is not well understood, but its effects on divers can be serious—drowsiness, dizziness, tremors, and nausea. Divers on the bottom who suffer HPNS symptoms obviously cannot work effectively, and in some situations they may be a hazard to themselves and others.

The current procedure for preventing HPNS on ultradeep dives is to pressurize the divers slowly (40 ft/hr according to the U.S. Navy tables) and then to give them saturation decompression at the end of the dive. In practical terms, a 1,000-ft dive would require 25 hr to pressurize the divers down to the bottom. If, for example, just 1 hr of bottom time were needed (as it might be on a drilling-rig support dive), the divers would be committed to 25 hr of pressurization (descent) and about 10 days of decompression time—a total of 11 days. This, of course, is an unacceptable price to pay for an hour of underwater work.

If HPNS could be eliminated and divers could be pressurized safely to the deeper depths within a matter of minutes, it is possible that relatively short decompression times could be used following 30- to 60-min work periods. Deep bounce diving on drilling rigs would then be practical. The alternatives, until more research is conducted, are saturation diving procedures, unpressurized (one-atmosphere) diving systems, and remotely operated vehicles for diving support on drilling rigs working in water depths beyond 500 ft.

Bounce diving to 1,000 ft is not one of the industry's major problems. The biggest concern today is depths ranging from 500–800 ft. This range is where much of the deeper diving activity will occur during the next few years. Short-duration bounce diving at these depths will require using decompression schedules that are not well tested and that may produce an unacceptable incidence of bends. Therefore, contracts calling for this kind of capability, such as deep drilling rig diving, must be supported with more sophisticated diver life-support systems, monitoring instrumentation, and environmental control equipment than has been necessary at shallower depths.

SUMMARY

Today the basic physiological constraints in diving, at least in the depth range in which divers usually are needed to support the offshore oil industry, are well understood. Safe operating procedures and new equipment have been developed that virtually eliminate inert gas narcosis, decompression sickness, and other less-publicized problems such as oxygen toxicity, excessive CO_2 buildup (hypercapnia), and low-temperature exposure (hypothermia). However, as the oil industry continues to move into increasingly deeper water, there is no question that new research in diving physiology and additional new equipment development will be needed.

Not many years ago, diving industry leaders were confident that divers would be working on the sea bed in 1,500-ft (and deeper) depths in the 1980s. Many of the experimental dives that were designed to prove this concept actually did just the opposite—they clearly showed there were some significant problems related to ultradeep diving. In addition, the few deep working dives that were conducted by the diving service companies in the field indicated that 1,000-ft work would be prohibitively expensive because of the necessary technical complexity of the diving systems, the need for large specialized surface-support crews, and the high consumption rate of costly breathing gas mixtures. The awareness of the difficulties and costs involved in ultradeep commercial diving led to the renewed interest and recent development of practical unpressurized, one-atmosphere diving systems (such as the ARMS observation/manipulator bells and JIM and WASP units) and remotely operated vehicles (i.e., TROVs, TRECs, RECONs, RCVs, etc.) for deep diving work.

The development of these underwater work systems is bringing interesting changes in the commercial offshore oil-field diving business. One-atmosphere and remotely operated systems, although primarily designed for deep-water use, can sometimes be used effectively in shallow water, thereby displacing the in-water diver. On the other hand, the use of the new systems by the diving service companies has begun to broaden the overall scopes of their businesses. The systems are now being used for certain kinds of survey, inspection, and deep drilling rig support work that would have been impossible for the diving companies to obtain with conventional diving service capabilities only.

18 Diving Capabilities and Equipment

Diving capabilities, from an equipment standpoint, can be categorized broadly as air diving and mixed-gas (helium-oxygen) diving. Air diving is the usual method for performing work in water depths of less than 170–200 feet. Beyond that depth the probability of the diver being afflicted with nitrogen narcosis precludes the use of air as a breathing medium. Virtually all diving beyond 200 feet is now done using mixed-gas equipment.

All diving equipment falls into three main categories: self-contained equipment (e.g., scuba gear), surface-supplied equipment (air diving systems and mixed-gas diving systems), and bell diving systems (bounce and saturation bell diving complexes for deep and long duration diving which utilize mixed-gas).

AIR DIVING

Self-Contained Underwater Breathing Apparatus (Scuba). Scuba equipment utilizes high-pressure cylinders to store compressed breathing air. The air is supplied by a demand breathing regulator that delivers the air to the diver at a pressure equal to the surrounding water pressure. The diver's exhalation is exhausted into the water.

The advantages of scuba are portability, because of its small size and light weight; horizontal mobility, because the system does not require any connections to the surface; and a high degree of depth flexibility and buoyancy control. The disadvantages of the system are a depth limitation of 100–130 ft under ideal conditions, a limited gas supply, and an exertion limitation, caused by the breathing resistance inherent in the demand regulator. A serious operational disadvantage is that there is usually no communication between the scuba diver and topside personnel. Scuba is seldom used in commercial diving operations because most underwater work requires an unlimited breathing gas supply, minimal breathing resistance, good two-way communications for efficiency and safety, and little degree of horizontal mobility or topside portability.

Surface-Supplied or Lightweight Equipment. Surface-supplied or lightweight commercial diving equipment falls into two general categories. The first can be called face mask gear, which incorporates a full face mask supplied by an air hose from the surface. The mask usually includes a communication system. The most commonly used full face mask systems include an optional oral-nasal demand as well as free-flow breathing, such as the Kirby-Morgan band mask. With full face mask equipment, the diver generally wears a standard neoprene wet suit, a lightweight dry suit, or a hot-water suit.

The second category is helmet equipment, which is comprised of a surface-supplied lightweight fiberglass or metal helmet that is functionally similar to a hard hat. The helmet also incorporates two-way communications, as does almost all surface-supplied commercial diving equipment. Helmet systems may be used with neoprene wet suits (wet dress) or with various types of dry suits or hot-water suits. Both systems include weight belts and boots or swim fins. Bail-out bottles are used with lightweight gear when diving to depths greater than 100 FSW (feet of sea water) and when bell diving. Lightweight helmet gear offers an unlimited air supply, good communications, good mobility, and minimum physical restrictions. Helmet gear also provides head protection and, when used with a dry suit, offers buoyancy control and effective protection in cold water.

The necessary support equipment for surface-supplied lightweight full face mask and helmet gear includes an air compressor and volume tank, an air manifold system, communications equipment, a diving stage, and, if the job is deeper than 70 ft or is outside the no-decompression limits, a decompression chamber. A class II open diving bell often replaces the diving stage of deeper surface-supplied air jobs as a diver transport method. The open bell can be used as a bottom refuge in case of a diver equipment malfunction underwater. Air diving with surface-supplied lightweight equipment is usually limited to 190-200 ft because of the possibility of nitrogen narcosis.

Deep-Sea (Hard-Hat Equipment). Deep-sea or hard-hat gear consists of a metal hard hat, a watertight suit (dry dress), a heavyweight belt, and ankle weights or weighted boots. As with lightweight gear, an umbilical from the surface provides the air supply and communications. The advantages of deep-sea gear are unlimited air supply, maximum physical protection, good working leverage (because the divers have a wide buoyancy control range and can make themselves very heavy underwater), and good temperature protection. The disadvantages of the equipment are its bulk and excessive air (surface) weight. The depth restrictions and surface-support equipment required for hard-hat equipment are essentially the same as for lightweight gear. Hard-hat gear gradually is being phased out of the offshore oil-field diving business. In most areas of the world, the equipment has been replaced by lightweight helmet gear.

MIXED-GAS (HELIUM/OXYGEN) DIVING

Mixed-gas diving equipment is essentially the same as air diving gear except that some systems incorporate a means of partially recirculating (rebreathing) the expensive premixed helium/oxygen gas to avoid exhausting all of it into the water. Mixed gas is also used sometimes with full face mask or helmet oral-nasal equipment operated in the demand open-circuit breathing mode.

In addition to the usual surface-supplied air diving support equipment, mixed-gas diving requires a manifold or regulator console (rack box) that lets surface personnel switch the diver's breathing supply from air to the appropriate helium-oxygen mixtures and back to air during descents and ascents. In most situations, a helium voice unscrambler that makes the mixed-gas diver's speech intelligible to surface personnel is also used. An on-site decompression chamber is always required for mixed-gas diving.

BELL DIVING SYSTEMS

At this time there are probably well over 300 closed diving bell systems in operation around the world. They are almost always combined with deck decompression chambers (DDCs) to which they can be mated after a dive to let the diving bell occupants transfer to a larger, more comfortable pressure vessel. Job conditions and water depths generally dictate when bell systems should be used. As a rule, bell systems are highly desirable from a diver efficiency and safety standpoint in depths beyond 200 ft. Bell diving systems are used on drilling-rig support diving operations from floaters (drillships and semisubmersibles) working in water depths in excess of 200 ft and on offshore construction and inspection operations requiring saturation diving.

Bell systems with DDCs capable of supporting saturation diving operations are necessary when bounce diving depths exceed about 400 ft and when back-to-back bounce diving may be necessary because of the possible need for saturation decompression. Saturation bell diving systems with large multiple DDCs always support offshore construction and inspection diving programs where the divers must live under pressure for long periods to complete the work.

Diving Bell

The diving bell is a pressure vessel designed to be mated by a quick connect-disconnect coupling (e.g., a tube turn) to the deck decompression chamber complex, allowing diver transfer under pressure between the diving bell and the deck decompression complex. It provides a safe, comfortable means of transporting divers under pressure to and from the bottom work site. The bell atmosphere is maintained free of CO_2 with an internal chemical scrubber system, and it is usually heated by a hot-water heat exchanger system.

A bell may serve as an observation chamber when used at one atmosphere. In this mode it can conduct visual or TV surveys of the work site prior to pressurizing the divers. This procedure can shorten the diver's bottom time on bounce dives significantly because everything related to doing the work can be organized and prepared on the bottom before pressurization begins. A lift line (load line) and umbilical connects the diving bell to the surface. The umbilical provides power, communications, gas supplies, and hot water to the bell from the surface. The divers themselves are tethered to the bell with umbilicals that provide similar services from the bell out to the working diver.

Deck Decompression Complex

The deck decompression complex usually consists of two or more pressure chambers that provide safe, comfortable living quarters for the divers while they are under pressure between dives or while they are decompressing. The number and size of the deck chambers depend on the scope of the operation, the number of divers, the duration of their stay at bottom pressure, and their anticipated decompression requirement. The chambers that comprise the deck decompression complex usually are designed with bolt flanges so they can be bolted together in different configurations to meet varied operational and space requirements.

Power, communications, gas distribution, and gas monitoring services are provided to the deck decompression complex (and the diving bell) by means of electrical cables or flexible hoses. These services are controlled and monitored through a central console housed in a separate enclosure commonly called a control van.

Control Van

Control and monitoring of all functions and services of the bell and deck decompression complex are maintained by the equipment and instruments housed in the control van. The control van provides weather and noise shelter for the surface support crew and usually is designed for safe operation within the designated explosive atmosphere area on exploratory drilling vessels. Most of the auxiliary and support-system controls and monitors are routed to the diving bell and deck decompression complex through the control van.

Auxiliary and Support Systems

The auxiliary and support systems commonly used with commercial bell diving systems include the following.

Gas Storage Systems. The deck decompression chamber in which bell divers live during their stay under pressure usually contains a mixed-gas (helium-oxygen) atmosphere. The pure or premixed supply gases are stored in high-pressure cylinders manifolded together into supply banks. These are connected to the appropriate control van gas distribution panel so the gas can be routed to the various deck decompression chambers and the bell.

Gas Booster Compressors. High-pressure gas booster compressors replenish the supply banks by pumping lower-pressure gas from a partially depleted bank into the high-pressure supply bank. They also produce different mixtures from pure or premixed gas supplies.

Gas Mixers. On-line gas mixers simplify logistic support of diving operations by providing an on-site means of producing various gas mixes from pure oxygen, helium, or nitrogen supplies. Gas mixers eliminate the need to store large amounts of premixed gas, some of which might not be needed on a specific diving operation.

Air Systems. The source of compressed air for diving systems usually comes from low-pressure (200–300-psi) air compressors. High-pressure (3,500–5,000-psi) air compressors are sometimes used as supplementary air supply sources.

Environmental Control Units. Environmental control units (ECUs) circulate the deck decompression chamber atmosphere, automatically adding makeup oxygen and removing CO_2 and odors. ECUs can also dehumidify and heat or cool the chamber atmosphere. Most ECUs are comprised of modular internal chamber circulation systems and external conditioning units. Backup ECUs usually are plumbed into each individual chamber with crossover piping systems.

Heating Systems. Hot-water diver heating systems provide heat to the surface-supplied diver and to the diver working outside the diving bell, preheat the divers' breathing gas, and heat the interior of the diving bell and the deck decompression complex. Hot-water diver heating systems are usually fuel (diesel) fired or steam supplied from the onboard ship, rig, or barge steam supply.

Helium Voice Unscramblers. The density and acoustic properties of the helium-oxygen gas mixes that divers breathe in deeper water have a peculiar "Donald Duck" effect on their speech. The greater the depth, the greater the percentage of helium and the greater the voice distortion. Electronic unscramblers substantially reduce the distortion and make the divers' speech intelligible. Unscramblers are used on virtually all mixed-gas diving operations in the offshore oil-field diving industry.

Bell Handling Systems. Launching, recovering, and mating a diving bell to a deck chamber requires a complex handling or lift system designed and configured for the particular diving system and for the operational conditions under which it will be used. The handling system must also be compatible with the particular vessel, barge, or drilling rig on which the diving system will be installed.

Handling systems may be built around off-the-shelf winch or crane designs or they may be of special design. They may be self-contained portable units or an integral part of the support structure of the surface support vessel or rig.

A hydraulic winch system generally handles the diving bell load line. The bell umbilical usually is handled by a separate tuna-block-type winch. If the bell umbilical is handled manually, it is usually secured to the load line with chain stoppers so the load line carries most of the umbilical weight.

Bell Diving System Installation

Commercial bell diving systems are as portable and adaptable to different support vessels as possible. But often, particularly with larger, deeper rated systems, considerable design effort may be involved in interfacing the diving system with a particular vessel. Regulatory agencies such as the USCG, ABS, Det Norske Veritas, and Lloyd's generally require certified diving systems drawings and interface engineering submissions prior to the actual inspection of the vessel/diving system installation. The diving system interface with the vessel can be divided into four categories.

Structural Interface. This includes placement of the diving system, auxiliary support equipment, gas supplies, and the handling system on the vessel's deck. The structural interface between the vessel and the diving system can present problems from an engineering point of view, usually because of weight restrictions. At this phase of the program, all aspects of the diving system, its associated auxiliary equipment, and the gas storage system must be considered. The deck strength and handling system structural interface usually will pose the most complex engineering considerations. Some drilling companies are now designing diving system requirements into the basic structural designs on new specialty diving vessels and drilling rigs.

Electrical Interface. In the past few years the electrical interface between the diving system and the vessel or rig has become more complex. One of the reasons is the explosion-proof criteria for diving systems on drilling rigs. This, coupled with the increased electrical power requirements of the deeper diving systems, makes electrical interface a major consideration on most diving system installations. A typical rig diving system will have the following electrical requirements:

1) handling system electrohydraulic winch:
   125 hp, 440 v
2) LP compressor:
   30 hp, 440 v
3) diving system control van:
   20 hp, 220 v
4) environmental control system:
   20 hp, 440 v
5) auxiliary equipment van:
   15 hp, 440 v
6) miscellaneous equipment, lighting, etc.
   15 hp, 220/110 v

Gas and Fluid Systems Interface. The requirements for these services are generally minor in nature. A diving system requires the vessel's compressed-air service for air tuggers and hot-and-cold potable water for sanitary services. Steam is often needed for heating the deck chamber complex and the bell diver heater units on drilling-rig installations. Adequate steam is not available on some rigs, and the quantities required for the diving system (when it is in use) can be considerable. Self-contained, fuel-fired heaters can be used on rigs that cannot provide adequate steam service, but these heaters must be located outside the rig's explosive-atmosphere area. Diving system heating on construction barges and specialty diving vessels where explosion-proof equipment usually is not required is provided by self-contained, fuel-fired heating units.

Handling System Interface. The bell handling system presents the most complex aspect of the diving system/vessel interface. Once the deck chamber complex location is determined, the handling system must be installed so the bell can be launched and recovered efficiently. Often, the configuration of the vessel will rule out standard handling system designs because of overhead interference, excessive distance for bell translation, or permanent vessel equipment obstructions. For this reason many bell-handling systems must be custom-designed for installation on a specific vessel.

The most common types of bell handling systems are 1) articulating A or H frame, 2) overhead trolley (gantry crane), 3) bell rotational system, and 4) combination bell rotational system and trolley.

The articulating A or H frame is utilized when the deck chamber complex can be placed very near the bell deployment site. The frame translates the bell from the mated to the launching position with hydraulic rams.

The overhead trolley is employed when the bell deployment site is 12–30 ft from the mating position. The bell is lifted to the trolley and locked into position. The trolley then translates the bell horizontally to the launching or mating position.

The bell rotational system rolls a single-hatch bell 90° to a horizontal position for side mating with the deck chamber complex. This is accomplished with trunions on the bell frame that impart a high overturning moment when the bell is landed on a flat surface. The system is employed when the deck chamber complex is close to the deployment site. This system decreases the overall height of the diving system installation because the single-hatch bell is side-mated rather than top-mated to the deck chamber complex.

The combination bell rotational frame and trolley is employed on vessels where a rolling door covers the diving moonpool. The bell is landed on the rolling door and rotated 90° to horizontal. The rolling door then translates the bell to the mating position with the deck chamber complex.

COMMERCIAL SUBMARINES

The commercial submarines or submersibles in offshore oil fields are either one-atmosphere observation subs or combined one-atmosphere/diver lock-out subs. Each type usually carries a hydraulic manipulator system that lets certain kinds of work be performed without using divers. Commercial subs are self-contained, one-atmosphere bells or diver lock-out bells with built-in buoyancy control systems, horizontal propulsion capability, and no attachments to the surface.

Observation subs are used primarily for the various types of offshore inspection work that require extended horizontal mobility, such as underwater pipeline route surveys and installed pipeline inspections. Diver lock-out subs are used primarily when divers must enter the water, such as inspections of the interior structures of steel platforms where it would be hazardous to use conventional diving bells, some types of routine construction work, and emergency underwater repair work. In the latter category diver lock-out subs have proven very effective in the field.

The disadvantages of commercial subs are (1) the relatively high cost of operations, (2) the limited onboard diver breathing gas supply on lock-out subs, and, on jobs where long horizontal transits or extensive diver heating are required, (3) the limited onboard power supply. Practical development work is being done on closed-circuit self-contained and push-pull diver breathing systems and more efficient sub battery systems. These new systems may eliminate the limited gas supply and power disadvantages of lock-out subs.

The first commercial submarine lock-out dive for the offshore oil industry was done in the Gulf of Mexico by Oceaneering International in 1967. There was little subsequent activity until 1975 when subs began to be used in the North Sea for inspection and construction support diving work. In 1977 and 1978 Ocean Systems' diving crews working from two NOL/Intersub lock-out sub spreads (*Intersub III* and *Intersub IV*) logged more than 20,000 manhours in saturation on construction-related jobs. Today subs are being used for many types of inspection work, including NDT and buried pipe tracking, as well as for subsea construction support and emergency repair work.

UNPRESSURIZED (ONE-ATMOSPHERE) DIVING SYSTEMS

One-Atmosphere Observation/Manipulator Bells. One-atmosphere observation/manipulator (O/M) bells were first introduced to the offshore oilfield diving industry by Oceaneering International (USA) and Comex (France) in 1976. They were designed primarily to provide underwater work capability to support exploratory drilling-rig operations in very deep water. There are about six or eight O/M bells in the field today, some of which are being used on rigs working in depth up to 3,000 ft.

O/M bells let operators work in deep water in an unpressurized, shirt-sleeve environment. They therefore need not be decompressed after a dive, regardless of the length of time underwater. The work is accomplished with visual and TV observation and sophisticated manipulator systems.

The main advantage of O/M bells, compared to conventional deep diving systems, are extreme depth capabilities, unlimited repetitive dive capability, less complex and lighter surface support equipment spreads, and smaller crew requirements. The disadvantages of O/M bells are restricted horizontal mobility, because they are tethered to the surface by a load line/umbilical; restricted work capability, because everything must be done with manipulators; and, on earlier models, restricted vertical mobility at the underwater work site, because of the lack of built-in buoyancy control systems.

Manned Articulated Diving Suits and Capsules. The one-atmosphere articulated diving suit was introduced to the offshore oil-field diving industry by Oceaneering International in 1974 as the 1,000-ft rated JIM suit. This system and the newer versions that are rated to 1,500 and 3,000 ft are individual pressure-resistant magnesium alloy diving suits that have self-contained life-support systems. They are tethered to the surface with a lift wire and communications cable. So far, the suits have been used for deep-water recovery work and exploratory drilling rig support diving in depths of more than 2,400 ft. Two suits, one of which is used as a backup system, are deployed on a job. At present, Oceaneering has 18 operational one-atmosphere, JIM-type diving suits.

The advantages of one-atmosphere diving suits are extreme depth capability, virtually unlimited repetitive dive capability, and relatively small equipment and surface crew requirements. The disadvantages are restricted underwater work capability because the fingers or hand manipulators are manually operated and restricted vertical and horizontal mobility.

A small number of recently developed one-atmosphere capsules or partially rigid suits like the 2,000-ft rated WASP system are similar in capability to the JIM suit. These tethered units have articulated arm sections and rigid lower sections, and they can be propelled horizontally and vertically by electric thrusters. The devices are being used for platform inspections and other types of subsea observation and inspection jobs.

UNMANNED TETHERED REMOTELY OPERATED VEHICLES

Remotely operated vehicles (ROVs) are basically tethered underwater TV camera platforms with hydraulic manipulator systems and electric or electrohydraulic propulsion systems (Fig. 18–1). The units are controlled from the surface with TV systems for operator observation and joy stick vehicle propulsion and manipulator controls. Some of the larger vehicles can carry inspection, instrumentation, and tool payloads of more than 300 lb and have depth ratings of 3,000 ft. The major builders of remotely operated submersible vehicles are International Submarine Engineering of Vancouver (TROV, TREC, TRUC, and DART), Hydro Products of San Diego (RCV), and Perry Submarine Builders of Riviera Beach, Florida (RECON).

The development of remotely operated vehicles over the last few years is one of the most important recent technological advances in the commercial

FIG. 18–1 One of the world's deepest-rated ROVs is this TROV by International Submarine Engineering, Vancouver. Soluc Ocean Systems operated the TROV on a 3,000-ft deep pipeline route survey in the Mediterranean. *(Photo by Chuck Collins)*

diving business. These vehicles are being used routinely on inspection and NDT jobs and, in some cases, on light manipulator work jobs. Originally, the vehicles were designed for extremely deep diving and to provide television inspection in places that were inaccessible to conventional divers. In the past five years, remotely operated vehicles have proven reliable in U.S., North Sea, and Latin American field operations. Their utilization is expected to increase significantly in the next decade. They may well be used routinely in shallow water less than 600 ft because they are relatively inexpensive to operate. However, at present the vehicles usually are considered primarily TV observation, inspection, and light-manipulator tools with ultradeep water capability.

SPECIALTY DIVING VESSELS

Specialty diving vessels are used mainly for offshore construction and inspection work. The larger vessels are usually owned and operated by the oil companies. The smaller vessels are owned or chartered by diving companies and are designed to compete with the large construction barges for offshore jobs that do not require heavy lift capability. Most specialty diving vessels are now being used in the North Sea and Latin American areas. They can be put into three broad categories, each of which has a definite specialty: (1) multipurpose service (or support) vessels (MSVs), (2) construction/diving vessels (CDVs), and (3) diving support vessels (DSVs).

Multipurpose Service (or Support) Vessels

MSVs are large, oil-company-owned, self-propelled semisubmersibles (or, in at least one case, a modified ship) that provide offshore construction, inspection, and safety-emergency services, such as fire fighting, pollution control, and hospital accommodations. They are designed as year-round North Sea safety vessels but are used primarily in the offshore areas being developed by the vessel owners. Diving personnel and equipment on MSVs usually are provided by subcontracted diving service companies.

Most MSVs are dynamically positioned to enable them to operate at any work site without risk of anchor or anchor cable damage to underwater structures. Semisubmersible MSVs are ideal for offshore construction and inspection work because of their stability (e.g., estimated North Sea work time is up to 300 days/year), dynamic positioning capability, and extensive accommodation, storage and work space. However, MSVs are extremely expensive to acquire, maintain, and operate, which is why they are owned by oil-company consortiums rather than by the offshore construction or diving companies.

Construction/Diving Vessels

CDVs are large ships that have been modified to perform offshore construction and diving work. They have capabilities similar to conventional derrick or utility barges owned by the offshore construction companies, but most CDVs are smaller and have dynamic positioning systems, integrated saturation diving systems, and firefighting capabilities (Fig. 18–2). Some can support underwater pipeline welding operations. They generally do not have the heavy lift capability or the deck space of conventional construction barges. CDVs were introduced by diving companies to compete with offshore construction company barges and, in some cases, to serve as temporary field safety vessels in the North Sea until the oil companies took delivery of their MSVs.

CDVs usually are products of a joint venture between a vessel owner and a diving company. The diving company is responsible for operating the vessel in its working mode and for marketing the total package to the oil companies.

The main advantage CDVs have over MSVs or conventional construction barges is lower operating cost. The DP-equipped CDVs can also work in areas where barges cannot anchor safely. The disadvantages of CDVs in relation to barges are limited lift capability, inferior stability, and, in most cases, less working deck space.

Diving Support Vessels

DSVs are usually converted monohull pipe carriers or large supply boats (220–260 ft LOA) that are dynamically positioned, have built-in diving systems with launching center wells, and 30–50-ton lift capabilities. Many have firefighting systems, so they can act as backup vessels in case of emergency. DSVs compete with conventional construction barges and CDVs for underwater inspection jobs, short and long-term maintenance work, light offshore construction jobs, and emergency repair work, predominantly in the North Sea. These vessels are also being used in Latin America and, to some extent, in other offshore areas around the world.

DSVs generally are obtained by diving companies through joint ventures or day-rate charter arrangements with supply vessel owners. Joint ventures are common when extensive modification of the vessel is done, i.e., DP systems, moonpool installations, extra accommodations, and heliports. Diving companies charter vessels on a day-rate basis only when they have short-term contracts where DP is not required and portable diving systems with over-the-side bell launching are adequate for the job.

FIG. 18–2 The *Samson Diver* is the largest ship-shape offshore diving/ construction vessel ever built. The ship, a converted oil tanker, was operated by Solus Ocean Systems in the North Sea. *(Photo by Larry Cushman)*

Monohull DSVs are relatively inexpensive to acquire and operate. Their main disadvantage in relation to the larger CDVs is inferior stability. Most DSVs cannot operate outside the 6–7-month North Sea summer weather window, although some of the vessels with well-designed bell handling systems and moonpools have achieved good performance during the winter months.

UNDERWATER WELDING

Underwater welding can be divided into four general categories: wet welding, hyperbaric or pressure-balanced dry habitat welding, one-atmosphere chamber welding, and other metallurgical fusion processes.

Wet Welding. Wet welding is the most common form of underwater metal joining. The process has been used for several decades and is similar to surface stick electrode (SMAW) welding. Wet welding, along with oxy-arc cutting, is taught to practically all divers (Fig. 18–3). In its simplest form, wet

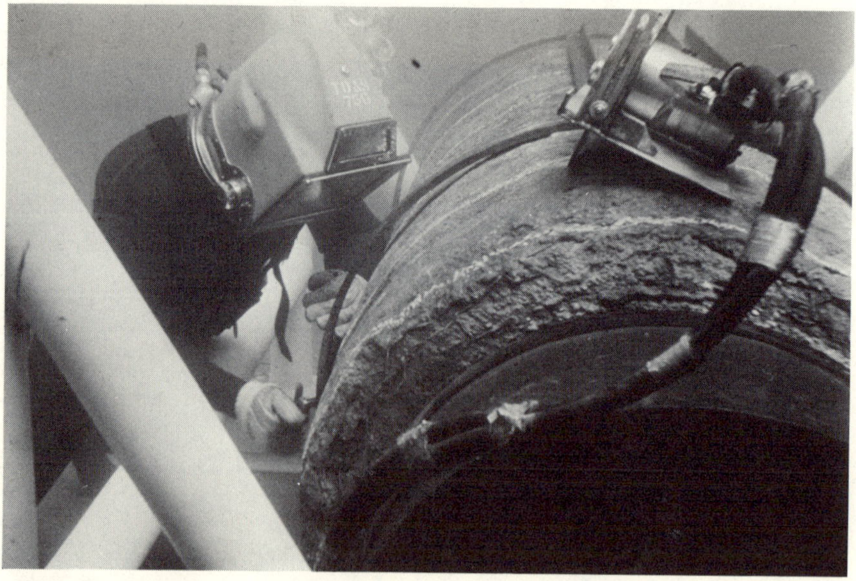

FIG. 18–3 This diver/welder is performing a wet-stick weld repair on a platform in the Gulf of Mexico. *(Courtesy Chicago Bridge and Iron/ SeaCon Services)*

welding employs a flux-coated ferritic electrode that has been waterproofed on the outer surface by a lacquer or similar coating. The rod is put in an electrode holder, provided with DC power, and dragged over the metal joint to be welded. The short-circuited rod melts to provide the weld deposit and fusion.

The advantage of wet welding is simplicity. However, by surface standards it usually does not provide a perfectly welded joint because of the presence of the surrounding water. The immediate quenching of the weld can result in hardening and reduced durability. Slag and gas are often captured in the weld metal, resulting in undesirable inclusions. Disassociation of the water and the uptake of hydrogen by the weld can result in embrittlement. Consequently, although the tensile strength of the weld may approach that achievable on the surface, durability and fracture properties are usually not comparable.

Although a number of new technical developments have been employed to improve wet weld quality, in its current form it usually cannot satisfy surface welding code requirements. Wet welding is most commonly used for structural repairs, anode attachments, seal welding of flanges, and other relatively low-stress jobs.

Hyperbaric Welding. Hyperbaric welding is done in an underwater habitat or enclosure filled with gas that is in pressure balance with the surrounding water. The objective is to exclude water from the vicinity of the weld to avoid the problems associated with wet welding (Fig. 18–4).

In its simplest form a hyperbaric weld may be performed in a small, lightweight enclosure (dry box) that covers the weld area and lets the diver-welders place their hands inside. More commonly, the enclosure or habitat is large enough for the diver to work inside around the pipe or structure to be welded. The simplest habitats are those used for making pipeline hot taps and structural repairs. These are essentially open-bottomed sheet metal boxes secured and sealed to the structure. For riser or pipeline tie-ins or pipe repairs, the habitat is usually within a structural framework that includes a hydraulic pipe alignment system that permits manipulating the pipe ends. Some hyperbaric welding systems employ additional separate hydraulic alignment units.

Because the fundamental problems imposed by water in the vicinity of the weld are overcome, hyperbaric welding can produce code-quality joints.

FIG. 18–4 A construction diver prepares to cut the concrete weight coating from a pipeline during preparations for an underwater dry hyperbaric tie-in weld. *(Courtesy Taylor Diving and Salvage)*

Pipeline welding normally is performed using TIG (GTAW) welding for the first two passes and stick (SMAW) for fill passes. Structural repairs are often done by semiautomatic wirefeed using flux core (FCAW) or stick welding.

Since the gas pressure in the habitat is the same as the pressure (depth) of the surrounding water, the welder encounters the same environmental and physiological constraints as a diver. Hyperbaric welder-divers are trained specially for this application. Normally, both the welder-divers and the specific welding process must be prequalified on actual pipe samples in a land-based hyperbaric pressure facility before the field operation is started.

Hyperbaric welding operations are costly because of the complex nature of the equipment, the number of specialized personnel, and the size of the surface support platform. However, hyperbaric welding is currently the only procedure available to weld underwater pipelines and pressure-vessel structures in accordance with the various API, Lloyd's, and DNV codes.

One-Atmosphere Chamber Welding

Since underwater hyperbaric welding imposes all of the complexities of a large saturation diving operation, development work has been done to perfect underwater welding systems that will let conventional surface welders inside function at one-atmosphere, i.e., surface, pressure. The welds produced under these conditions would be identical in quality to standard surface welds.

Two of the largest diving companies in the world have spent a great deal of time and money developing welding systems that will align pipelines, temporarily plug pipe ends, and enclose the weld area with a pressure-resistant, one-atmosphere work chamber. These systems must be pressure-sealed onto the equipment to be welded. Consequently they probably would not be suitable for some kinds of platform structural welding. For riser and pipeline tie-ins, such systems appear to offer cost advantages; although the welders would have to be transported back and forth in a one-atmosphere lock-on bell (or submarine), they would not have to be qualified as divers.

The practical development of this concept is a very complex technical task that has not been achieved successfully. Most pipeline alignment and sealing tasks currently performed by divers in hyperbaric underwater welding systems must be done remotely in one-atmosphere systems. Sealing one-atmosphere systems is difficult. Life-support equipment and habitat ventilation systems are more complicated. Whether this approach will be feasible technically and economically in the near future remains to be seen.

Other Metallurgical Fusion Processes

Other methods of fusing metal underwater are in development, and some have been used in the field. Most of these techniques are restricted to very specialized applications, such as anode attachment and small-diameter pipeline joining.

Cad welding uses a thermite chemical heat charge to fuse metals. The process has been used successfully underwater to attach anode straps to platform structures.

Explosive welding (metal fusion from high-energy shaped charges) is being developed by one diving company for underwater pipe joining. Basically, an explosive charge is placed inside the pipe joint to be closed, and an external collar is fitted over the joint. The charge is detonated, and the pipe metal is forced under extremely high pressure into the metal of the collar to form a fused joint. At this time explosive underwater welding has not been used successfully for pipeline tie-ins in the field.

A cold swaging process for joining pipelines underwater is presently being marketed in the industry. A hydraulically driven expanding and rolling internal mandrel is inserted into a pipe at the joint to be sealed, and an external collar is fitted over the joint. The mandrel is expanded and cold rolls the pipe into the collar to produce a homogeneous metallurgical bond. This process has been used successfully offshore to join risers to flow lines.

19 Future Trends

The major technological advances anticipated in the offshore oil-field diving industry over the next several years probably will be related to improving diver safety, reducing diver breathing gas consumption, and perfecting underwater hyperbaric welding processes, inspection (NDT) instrumentation, and remotely operated vehicle (ROV) systems. Little effort will be spent on ultradeep (over 2,000-ft) hyperbaric diving equipment development or physiological research. These are prohibitively expensive, there are significant problems, and demand does not yet exist to any great extent in the offshore oil industry.

Improvements in diver safety will result from introducing equipment that will monitor the divers' status and life functions while they are working underwater and from developing more sophisticated bounce and saturation diving field operating procedures. Other safety improvements will be related to operating diving bells from dynamically positioned vessels, using cross-hauled bells for interior inspections of production platforms, and developing improved bell and submarine emergency life-support systems.

The development of closed-circuit push-pull and self-contained mixed-gas diver breathing systems will continue, and these will become more commonplace in the field in the near future—particularly on deep saturation diving and submarine lock-out operations.

The technology of underwater welding will continue to advance at a rapid rate because most of the larger international diving companies will be trying hard to develop unique capabilities and to obtain a share of this high-revenue work. Underwater inspection and NDT instrumentation will also improve rapidly as the volume of this activity grows and becomes a larger part of the diving service business.

Remotely operated vehicles will be improved technically as their offshore use increases. Diving companies that cannot provide these systems will find themselves unable to bid certain construction and inspection jobs on which the vehicles are requested as part of the overall diving capability package.

Future business trends in the offshore oil-field diving industry are difficult to forecast, but three interrelated situations may combine to produce a trend toward fewer but larger international diving service companies. First is the high cost of new capital equipment items such as saturation construction diving systems, underwater habitat welding spreads, and specialty diving vessels. Smaller diving companies find it difficult to acquire the new systems necessary to support large-scale, long-term offshore construction and welding projects. Second, only companies with the financial resources to conduct research and development will remain leaders in the technically advanced areas of the industry. Third, more elaborate regulatory agency and insurance company requirements may make it difficult for some of the smaller diving companies to continue working offshore without upgrading their diving systems and operating procedures.

Another recent trend in this relatively new industry is toward more sophisticated operations management and engineering capability within the larger companies. This has been caused by the requirement to perform more complex underwater tasks in progressively deeper water and to perform new engineering-related tasks, e.g., code-quality welding, state-of-the-art testing, and of complex electronic and electrohydraulic equipment operation.

One of the most important side effects of these changes has been the significant reduction in the number of diving accidents over the last several years at a time when the number of underwater work hours around the world has increased steadily. This trend will continue.

Commercial offshore diving has grown out of its early pioneering stages. It is now a highly specialized, professionally managed, and technologically adept industry that provides a unique and vital service to the offshore oil industry.

REFERENCES

Cushman, Larry, and D. Michael Hughes. *Technology of Offshore Drilling, Completion and Production.* Tulsa: PennWell Publishing Company, 1976.

Cushman, Larry. "Diving Operations and Equipment." University of Texas/International Association of Drillling Contractors School of Offshore Operations presentation. Houston, 1974–1980.

Dugan, James. *Man Under the Sea.* New York: Macmillan Publishing Company, 1965.
Earls, Thomas. "Deep Diving from Oil Rigs—What are the Options?" Offshore Scotland Conference presentation. Aberdeen, 1973.
Gilman, Bruce C. "Design, Fabrication and Installation of Commercial Diving Systems." Equipment for the Working Diver Symposium, 1974.
Hughes, D. Michael. "Deep Diving Capabilities and Limitations." *Offshore,* 1973.
Hughes, D. Michael. "The Evolution of Commercial Diving Bells." *Oceanology,* March 1972.
Oceaneering International Inc. "Rig Diving Data" (unpublished document). 1972.
Solus Ocean Systems Inc. "Undersea Services Industry Analysis and Forecast" (unpublished document). 1978.
Somers, Lee H. *Research Diver's Manual.* Ann Arbor: University of Michigan Press, 1972.
Zinkowski, Nicholas B. *Commercial Oil-Field Diving.* Cambridge: Cornell Maritime Press Inc., 1978.

ABOUT THE CONTRIBUTORS

Michael W. Ridge

MICHAEL W. RIDGE is a member of the consulting staff of R. Stewart Hall. As a private consultant, Mr. Ridge has been associated with energy research projects in California, Washington, D.C., and Houston. He also spent several years in the Middle East in both administrative and managerial capacities. Mr. Ridge received his Bachelor of Science degree at Johns Hopkins University in Baltimore.

Ronald A. Weber

RONALD A. WEBER is manager of special studies in the Research and Development Group of Vetco Offshore Inc., Ventura, California. Mr. Weber's 18 years' experience in the design and application of offshore drilling and production equipment and systems includes serving as project manager for Exxon's Subsea Production System (SPS), the first major multiwell subsea production system. He also managed the initial design study for a subsea satellite production tree system for Shell Exploration and Production (United Kingdom) for the South Cormorant Field, North Sea, and was general manager of engineering for Vetco in the United Kingdom. Mr. Weber received his Bachelor of Science degree in mechanical engineering at Rice University, Houston, and is a registered professional engineer in Texas.

Larry Cushman

LARRY CUSHMAN is president of the Ocean Corporation, a Houston-based commercial diving school and specialized diving contractor engaged in underwater nondestructive testing and nuclear power plant repair work. Prior to joining the Ocean Corporation, Larry was vice president of corporate marketing for Solus Ocean Systems, an international diving contractor located in Houston. Before this assignment, he was in Aberdeen, Scotland, for three years as vice president and area manager of Solus Ocean Systems' Europe/Africa operations.

Larry has been in the offshore oilfield diving business since 1968 as a commercial diver, diver-trainer, manager and company officer. He worked for Oceaneering International for six years in various management positions, including marketing manager and president/director of the company's commercial diving school. He also worked for Perry Oceanographics as manager of their Ocean Products Division and was one of the founders of Sub-Marine Systems (now Kinergetics Inc.)

Charles Bridges

CHARLES BRIDGES, Gray Tool Co., has 21 years' experience in the design, development, and installation of many unique and ambitious offshore projects. He has a broad knowledge of the extreme variations in requirements for wellhead assemblies due to his design experience in equipment to be used on the North Slope of Alaska, the North Sea, the Middle East, Southeast Asia, and the Gulf of Mexico. He also has designed equipment and systems for extreme high-pressure applications.

C. Peterman

C. PETERMAN is general manager of the Advanced Ocean Systems Division of the Hydril Company. He has held this position since the founding of the division in 1978. Previous to this current assignment, Peterman held manager positions with the Valvcon and Control Systems divisions of Hydril.

Peterman graduated from Kansas University with a BSEE. He then served as a communications officer in the U.S. Airforce. Before joining Hydril, he worked for Shell Oil Company in general drilling and production assignments, followed by work on steamflooding and research and development work concerned with deep-water B.O.P. controls, wellhead reentry, and various production techniques.

F. Jeffrey Lane

F. JEFFREY LANE, Gray Tool, attended Texas Technology Institute and the University of Houston. His fourteen years with Gray Tool Company has included positions in wellhead and valve design, manager of valve design, and senior staff engineer. As manager of project engineering, he currently provides for technical coordination among Gray Tool engineering departments and other divisions of the company.

Mary Lou Moore

MARY LOU MOORE is the technical writer for Systems Research and Development, Vetco Offshore Inc., Ventura, California. In Tulsa, she performed in various writing capacities for Sunray-Midcontinent, Atlantic Richfield, Cities Service, Sinclair, and Skelly oil companies. She has served as a consultant to PPG Industries, Oldsmobile, Eastman Kodak, and American Parts Systems (a Gulf & Western Company). As a proposal development specialist for Rockwell International, Ms. Moore wrote technical documents regarding communications and navigation for many classified government programs. While at Rockwell, she was assigned to the president's staff, Electronic Systems Group, to write special technical and promotional material for all of the divisions of the Group. Ms. Moore attended the University of Tulsa and the University of Texas at Dallas, with emphasis in journalism and oil & gas law.

Index

Index